Rice in the Tropics

IADS DEVELOPMENT-ORIENTED
LITERATURE SERIES

Rice in the Tropics
was prepared under the auspices of the
International Agricultural Development Service
and the
International Rice Research Institute
with additional funding from the
German Agency for Technical Cooperation
(Deutsche Gesellschaft für Technische Zusammenarbeit)

Rice in the Tropics:
A Guide to the Development
of National Programs

Robert F. Chandler, Jr.

Routledge
Taylor & Francis Group

LONDON AND NEW YORK

First published 1979 by Westview Press

Published 2019 by Routledge
52 Vanderbilt Avenue, New York, NY 10017
2 Park Square, Milton Park, Abingdon, Oxon OX14 4RN

Routledge is an imprint of the Taylor & Francis Group, an informa business

Library of Congress Cataloging in Publication Data
Chandler, Robert Flint, 1907-
    Rice in the tropics.
    (IADS development-oriented literature series)
    Bibliography: p.
    Includes index.
    1. Rice. 2. Agriculture and state. 3. Rice—Tropics. I. Title. II. Series: United States.
International Agricultural Development Service. IADS development-oriented
literature series.
SB191.R5C523   338.1'8   78-19704

ISBN 13: 978-0-367-28605-7 (hbk)
ISBN 13: 978-0-367-30151-4 (pbk)

Contents

List of Tables ix
List of Figures xi
Foreword, by Sterling Wortman..................... xiii
Preface xvii

1. The Importance of Rice as a World Crop, and
   Its Principal Characteristics 1

   Area, Yield, and Production of Rice 1
   Rice as a Staple Food 9
   Types of Rice 12
   Quality Preferences among National Groups 15
   Rice as a Unique Food Crop.................... 17
   Kinds of Rice Culture 18
   Future Supplies of Rice 21

2. The Modern Rice Plant and the New Technology:
   Greater Potentials for Rice Production
   in the Tropics 31

   The Modern Tropical Rice Plant................. 32
   Response of Modern Varieties to Fertilizer........ 38
   Water Management in Lowland Rice 42
   Chemical Changes in Flooded Soils 43
   Solar Radiation and Rice Yields................. 44

Plant Protection 46
Mechanization for the Small Farmer 56
The New Technology and Farm Incomes 61

3. Problems of Postharvest Technology 65

Harvesting and Threshing 66
Cleaning and Drying........................... 68
Handling and Transportation................... 74
Storage 75
Rice Processing................................ 79
Rice Distribution and Quality Control 86
The Systems Approach 87

4. Rice Marketing 91

Local Marketing............................... 92
Self-sufficiency as a Goal 94
Export Marketing: Problems in Marketing
    Surplus Rice 96

5. Some Successful Rice Production Programs 101

Taiwan 103
South Korea................................... 113
The Philippines 123
Colombia 129
Other Countries That Have Made Rapid
    Progress 139
Comparing the Programs...................... 142

6. Promising Rice Research 145

Varietal Improvement 145
Supplying Nitrogen to the Rice Plant 153
Improved Insect Control at Low Cost 157
Better Weed Control Methods for Rainfed Rice ...158
Fundamental Causes of Low Rice Yields 159
Cropping Systems Involving Rice 160
Continuous Rice Production 162

7. Elements of a Successful Accelerated Rice
   Production Program 165

   Analyzing the Natural Resources 167
   The Essential Elements.......................171

8. A National Rice Program: Putting the
   Ingredients Together 189

   Organizing the Rural Structure 190
   Deciding Where To Put the Emphasis 193
   Prospects for Increasing Rice Yields in the
      Tropics....................................202
   Maintaining the Pace Once It Is Set 204

Appendix: Where and How To Get Assistance 207
Glossary...225
Annotated Bibliography............................229
Index 249

Tables

1. Average annual area, yield, and production of rice by region and country, 1961-65 and 1971-75 4
2. Apparent average annual per capita rice consumption in selected countries, 1971-75 8
3. Estimated maximum farm yields for 11 Asian countries, and the area of major types of rice land of varying yield potential 26
4. Area, yield, and production of paddy rice in 11 Asian countries in 1976, the weighted average of the calculated on-farm maximum yields (from Table 3) and the estimated amount of rice that could be produced in each country 28
5. Comparative labor requirements and costs of land preparation by power tillers and by water buffaloes in the Philippines in 1976 58
6. A comparison of the investment and operational costs of 6000-ton bag and bulk storage facilities...... 77
7. Average recovery efficiencies of three types of rice mills 83
8. Energy value and cost of four sources of energy in India 85
9. The economics of three sizes of milling operations 90

10. Government contributions and the total
    value of the completed projects in the New Village
    Movement in South Korea from 1971 to 1976 120
11. Average incomes of urban and rural families
    in South Korea from 1970 to 1975 121
12. Area, production, and yield of rice in Colombia
    from 1966 to 1976 (by sectors), and the
    percentage of the rice area under irrigation.........133
13. Proportion of rice area under lowland conditions,
    national rice yields, and yields of lowland and
    upland rice in 10 countries 143

Figures

1. Production, population, and per capita output of rice in the less developed countries, 1956-74 22
2. Resistance ratings of IRRI rice varieties in the Philippines 37
3. Effect of levels of nitrogen on grain yield of IR8, IR20, and Peta 39
4. Changes in pH of six soils after submergence 44
5. Grain yield of rice in relation to solar radiation during the 45-day period prior to harvest in 1968 45
6. The impact of insecticide placement on yield of transplanted rice 50
7. Yield response to low levels of nitrogen, with and without weed control 54
8. The relationship between the moisture content of paddy (IR8) at harvest to total field yield, the percentage milling yield, and percentage head rice yield 67
9. Chart showing the movement of paddy through a modern continuous-flow drying plant 72
10. Basic design of a modern rice mill 82
11. Sequence of postharvest operations 88
12. Export price of rice (Thai, 5 percent broken) FOB Bangkok, 1964-77 97
13. Indices of agricultural production, rice production, and population in Taiwan from 1950 to 1975 102

14. Yield of rice (paddy) in Taiwan, 1938 to 1975 103
15. Yield of rice (paddy) in South Korea, 1956 to 1976 . . . 114
16. Fertilizer consumption (NPK) in South Korea,
    1955 to 1975 115
17. Yield of rice (paddy) in the Philippines,
    1966 to 1976, compared with the 1961-65 average . . . 123
18. Yield of rice (paddy) in Colombia, 1966 to 1976,
    compared with the 1961-65 average 130
19. Average paddy yields in Colombia under irrigated
    and under upland conditions, 1955 to 1975 134
20. Average national yield of rice in Colombia in
    relation to percentage of rice land irrigated 135

Foreword

This book is of unusual significance. It is the first IADS-sponsored volume on a commodity written especially for authorities, nonagricultural as well as agricultural, in developing countries and for the assistance agencies which cooperate with them. To our knowledge, it is the first such volume on any food crop.

Its origins may be of interest.

In early 1976, representatives of IADS were invited by the government of a small Caribbean country to discuss ways to overcome low agricultural productivity and to alleviate the widespread poverty among its rural people. Food production was static. The population was increasing at high rates, as it had been for several decades. Demand for food was going up even more rapidly. And projections indicated that imports would escalate to levels dangerous for the country.

Rice was of particular interest: the planning unit of the ministry of agriculture had determined that national average yields of this basic commodity would need to be doubled within 10 years. Otherwise, massive imports requiring large outlays of scarce foreign exchange would be necessary. An effective research program had been under way for several years. It was clear that at least some of the technology necessary for a production push was available; the balance could be developed within a few years. We all agreed that a serious rice production campaign should be launched, with the goal of doubling national rice yields, from about 2.5 to 5.0 tons per hectare, in 10 years. Most of the increases would have to occur on rice farms that were a few hectares or less in size.

Recognition of the need for rapid increases in rice output was not difficult, for the planners had done their job well. The problem was to establish a plan for a nationwide rice program that political authorities of the country could understand and to present convincing evidence that the program could provide the necessary increases at reasonable cost. Government leaders were for the most part nonscientists and even nonagriculturists; their backgrounds were in law, business, education, or the military.

Unfortunately, we had no publication to give these authorities that would explain in clear language the technological basis for creating a national rice program tailored specifically to that country's needs. There was no shortage of scientific and technical literature. In fact, one could quite quickly get a computer printout of literally tens of thousands of articles on rice. But such arrays, important as they may be to scientists, are of little value to authorities in developing countries who need to know quickly the essential features of a commodity such as rice; what the essential components of a production campaign are; what successes or failures other countries have had, and with what approaches; what kinds of scientific and technical help are available from international centers and other countries or agencies; and what the world production and price situation is, or is likely to be. There was not a single publication we could recommend to the national authorities as the basis for design of an effective national rice program. This made discussions extremely difficult, even frustrating, for all concerned.

In scores of countries, rice is an important part of the economy. And in each country there are scores of individuals whose separate decisions can affect success or failure of national rice efforts. Consequently, IADS, in cooperation with the International Rice Research Institute, set out to produce a volume on rice that would marshal existing information on this crop, leaving out unnecessary detail, and present it in a form easily understandable to authorities in national governments or assistance agencies. The book was also to provide references to key items in the literature for those needing information in greater detail.

We turned, quite naturally, to Dr. Robert F. Chandler, Jr., the director of the International Rice Research Institute during its first 12 years. Probably no other individual has served as advisor to so many governments on the organization of national rice systems. Certainly no other individual has been decorated or otherwise honored by so many developing country institutions for contributions to their rice improvement programs. He has worked with the government of India on the establishment of its successful High Yielding Varieties Programme, and with the Philippines in laying the basis for the Masagana 99 rice production campaign. He has served as a consultant to the West African Rice Development Association, to the international banks, and to numerous other countries. His knowledge is both broad and deep. He has credibility.

During the course of the project, we found an ally in Deutsche Gesellschaft für Technische Zusammenarbeit (GTZ) of the Federal Republic of Germany. Their leaders, particularly Dr. Klaus Lampe, felt as we do that, given the tens of millions of dollars going into rice research annually, it is important that information be assembled and made available to developing countries in a form immediately useful to their authorities. We gratefully acknowledge the GTZ, The Rockefeller Foundation, and the Lilly Endowment for financial support and for encouragement.

Most developing nations have too few experienced agricultural leaders. Many of the individuals determining agricultural policy or managing agricultural programs have little knowledge about the crops and the agricultural production and marketing systems with which they must deal. Many want to be better informed. Most agricultural literature, however, is not suited for the nonexperts. Much of it is written in academic language by and for specialists. It is in fragments, scattered over a wide range of journals and other publications. One cannot obtain a comprehensive view without resorting to the study of many narrow articles in numerous publications. Moreover, much of the information is relevant primarily to temperate-climate regions, rather than to the tropics, where most developing countries are.

There are scores of commodities and problem areas for which comprehensive presentations of available information are needed, written in language understandable to both scientists and nonscientists. IADS is attempting to stimulate the production of publications on a wide range of subjects relative to agricultural development. They must be comprehensive, credible, and easily read. It is our hope that such books and reports will help officials of developing countries step up the pace of agricultural development, and point out where they can obtain additional sources of information and technical help.

Sterling Wortman, President
International Agricultural Development Service
New York City

Preface

Much has been written on the subject of agricultural development, on how to get agriculture moving, on the program required to persuade the small, subsistence farmer in the less developed countries to shift from traditional to modern methods of crop production. Yet, heretofore, that type of information as applied to a single major food crop has not been available in one publication.

Rice is by far the most important food crop in Asia, which contains over half the world's still expanding population. Thus it is through increased rice production that a large segment of mankind will continue to be nourished. Furthermore, it is on the rice industry and on the human employment it generates that much of Asia's rural economy depends. Rice, therefore, is likely to continue to play a primary role in the improvement of that economy. In Latin America and Africa, rice is important in a number of countries and promises to become more so as world demand for it increases.

Rice output must be doubled in the next 25 to 30 years simply to keep pace with predicted population growth. This can be achieved only by mounting accelerated production programs in those major rice-growing countries in which yields are now below 4 metric tons per hectare. Key people in starting and supporting such movements are the agricultural officials and the government planners and developers. It is for such busy administrators, who may not have time to read through numerous publications for the information they seek, that this book is primarily written.

Chapter 1 conveys the importance of rice as a world food crop and describes some of its unique qualities. In chapters 2 and 6 an effort is made to set forth the current stage of rice research and technology and to identify some of the more promising research projects that merit continuing attention. Chapters 3 and 4 treat postharvest problems, including ways of handling surplus rice production. Chapter 5 describes in some detail four national rice production efforts that achieved positive results. Chapter 7 contains a discussion of the more important elements of a successful rice production program, many of which are common to programs for the increased production of other crops. Chapter 8 is devoted to the subject of putting the elements discussed in chapter 7 into a workable scheme. The appendix describes the more important international agencies offering technical and financial assistance for agricultural development programs. Following the appendix are an annotated bibliography, arranged in the order of chapter content, and a glossary of terms.

I am grateful to the International Agricultural Development Service for inviting me to write this book and for providing financial support. Within that organization I am especially indebted to Francis C. Byrnes, who rendered helpful guidance while the manuscript was being written, and to Steven A. Breth, who provided much valuable advice during the review and revision stage of the writing and who expertly edited the final version of the manuscript.

The International Rice Research Institute generously supplied most of the photographs and made the final drawings of graphs and charts, many of which came from its own publications. The facilities of the excellent IRRI library were made available to me, and IRRI's director general and staff provided advice and services unstintingly.

The original draft of chapter 3 was written by James E. Wimberly and that of chapter 4 was prepared by J. Norman Efferson. Although the two chapters were revised somewhat, I attempted to retain the chief points made by each author. I am indebted to them both for their valuable contributions.

The original manuscript was reviewed by the following persons: M. S. Swaminathan, director general, Indian Council of Agricultural Research; Delane E. Welsch, professor of agricultural economics, University of Minnesota; Lloyd T. Evans, Division of Plant Industry, Commonwealth Scientific and Industrial Research Organization, Australia; Yujiro Hayami, professor of agricultural economics, University of Tokyo; T. T. Chang, geneticist, IRRI; S. K. De Datta, agronomist, IRRI; and Randolph Barker, agricultural economist, IRRI.

The comments of the reviewers were carefully considered and most of their suggestions were incorporated into the revised manuscript. I am obliged to each of them for his valuable ideas but assume full responsibility for any errors or omissions.

I am especially grateful to my wife, Muriel Boyd Chandler, who cheerfully and capably did the original editing and typing. Her suggestions and unflagging moral support were invaluable in preparing the book.

Robert F. Chandler, Jr.
Templeton, Massachusetts

1
The Importance of Rice as a World Crop, and Its Principal Characteristics

Rice is an ancient grain, the beginnings of its culture seemingly lost in prehistory. It is the staple food of approximately half of mankind. So dependent upon rice are the Asian countries that throughout history a failure of that crop has caused widespread famine and death.

Although the global production of wheat is greater than that of rice, about one-fourth of the wheat crop is used for nonfood purposes compared with only 7 percent of the rice crop. Furthermore, rice is far more important than wheat in the less developed countries of Asia, where over half of the world's population lives. In China, with its 900 million inhabitants, for instance, the consumption of rice is 2.5 times that of wheat, the second most important food crop of that country. A similar example is India, where a population of over 600 million persons consumes more than twice as much rice as wheat.

Except of course for Antarctica, every continent on the planet produces rice. It is grown from the equator to latitudes of 53 degrees north (in China) and 35 to 40 degrees south and to elevations (in tropical regions) as high as 2400 meters above sea level. Its importance as a food crop expands as man's numbers increase—the crucial problem at this stage being whether rice production can manage to keep pace with population growth until the latter comes to a long overdue halt. Meanwhile, the need intensifies for maximum knowledge of this vital plant.

Area, Yield, and Production of Rice

In Table 1, statistics for unhulled rice, called rough rice or

1

paddy, are given for two contrasting periods. One is the 5-year base period 1961-65, when yields were rather constant and before the popularly termed "Green Revolution" had begun. The other is 1971-75, which includes the drought year of 1972 as well as the generally good years 1973 and 1975. The data for 1971-75 reflect the progress that followed the introduction of modern rice varieties on a large scale and the increased use of fertilizer and other inputs. To convey the changes that took place in the first 10 years after modern rice varieties were introduced into the tropics and subtropics, the table also gives increases in area, yield, and production expressed in percentages.

Asia

Table 1 shows that Asia produces over 90 percent of all the rice grown. For that reason, the average rice yields for the world and for Asia are essentially the same, as are the percentage increases in rice area, yield, and production. The average percentage of increase in yield in Asia during the 10-year period was about double that of the increase in land area devoted to rice. In most Asian countries, a large proportion of the land suitable for the cultivation of rice is already being used for that purpose. In contrast, in Africa and South America, where vast tracts of arable land still remain uncultivated, there has been little change in yield. Production increases have been due almost entirely to expansion in the area planted to rice.

The 20 Asian countries that plant over 200,000 hectares of land to rice can be divided into three broad categories:

1. Countries obtaining high average yields (over 3 t/ha) and having a potential for at least moderate yield increases; this group includes China, Iran, Japan, North Korea (Democratic People's Republic of Korea), South Korea (Republic of Korea), and Taiwan
2. Countries with low to medium yields (1.5 to 2.5 t/ha) but showing consistent yield increases from year to year (above 15 percent between 1961-65 and 1971-75): Pakistan, Indonesia, Philippines, Afghanistan, Malaysia, and India

3. Countries with low yields (less than 2 t/ha) or showing low levels (less than 15 percent) of increase from year to year; in this category are Vietnam, Sri Lanka, Laos (included here, in spite of its rapid increase in yields, for reasons later explained), Kampuchea, Burma, Bangladesh, Thailand, and Nepal

Within each category the countries are similar in the proportion of rice land being irrigated and the amount of chemical fertilizer being used. For the countries in category 1 over 95 percent of the rice crop is irrigated. The corresponding figure for the nations in category 2 is 40 percent, and for those in category 3, 25 percent. Actually, the statistics exaggerate the amount of rice land being irrigated for categories 2 and 3, because they include not only the areas with irrigation systems that permit year-round cropping, but also land that has irrigation only as a rainy season supplement. For example, the Philippines is listed in Food and Agriculture Organization (FAO) reports as growing 40 percent of its rice crop under irrigated conditions, but in fact only 15 percent of the crop is grown in fields supplied with sufficient water to provide for a second crop during the dry season.

Unfortunately, there are no accurate data on the amount of fertilizer applied to the rice crop alone. However, a fair indication can be obtained by examining the quantity of fertilizer used per hectare of arable land. In 1973-74, the countries in category 1 used 275 kg/ha of nutrients (nitrogen, phosphorus, and potassium), while countries in category 2 used only 31 kg/ha, and those in category 3, only 20 kg/ha.

Of course, irrigation and fertilizer use are not the only influences on yield levels and the rate of annual progress in improving them. War conditions and political disturbances, for instance, definitely have hindered progress in Laos, Kampuchea, and Vietnam.

All six countries in category 2 (those maintaining more than a 15 percent gain in yield) have mounted strong rice production efforts. As a result, percentage increases in both yield and total production for those nations are greater than the average for all of Asia.

TABLE 1. AVERAGE ANNUAL AREA, YIELD, AND PRODUCTION OF RICE BY REGION AND COUNTRY,[a] 1961-65 AND 1971-75

Region or Country	1961-65			1971-75			Change (%)		
	Area (thousand ha)	Yield (t/ha)	Production (million tons)	Area (thousand ha)	Yield (t/ha)	Production (million tons)	Area	Yield	Production
ASIA									
Afghanistan	214	1.60	0.34	204	1.97	0.40	-4.6	22.7	17.2
Bangladesh	8955	1.68	15.03	9737	1.71	16.96	8.7	2.0	12.8
Burma	4741	1.64	7.79	4840	1.73	8.40	2.1	5.6	7.9
China	30,180	2.75	83.10	34,137	3.17	108.34	13.1	15.2	30.4
India	35,626	1.48	52.73	37,460	1.72	64.11	5.1	16.3	21.5
Indonesia	7036	2.04	14.38	8326	2.54	21.17	18.3	24.4	47.2
Iran	276	3.08	0.85	371	3.28	1.23	34.4	6.4	44.6
Japan	3281	5.01	16.44	2690	5.83	15.67	-18.0	16.3	-4.6
Kampuchea	2284	1.08	2.46	1054	1.20	1.40	-53.8	11.1	-43.2
Korea (DPR)	622	3.99	2.48	701	4.63	3.26	12.7	16.1	31.5
Korea (Rep.)	1169	4.11	4.81	1207	4.89	5.90	3.2	18.8	22.7
Laos	728	0.84	0.61	672	1.29	0.87	-7.6	53.9	42.2
Malaysia	535	2.13	1.14	760	2.57	1.95	42.0	20.6	71.4
Nepal	1099	1.95	2.15	1256	1.95	2.45	14.3	0.0	14.3
Pakistan	1287	1.42	1.82	1553	2.31	3.58	20.6	62.9	96.5

(Table 1 cont.)

Philippines	3147	1.26	3.96	3451	1.59	5.48	9.6	26.3	38.5
Sri Lanka	505	1.91	0.97	635	2.13	1.36	25.7	11.4	40.1
Taiwan	773	3.80	2.94	757	4.28	3.24	-2.0	12.7	10.3
Thailand	6026	1.72	10.38	7467	1.87	13.95	23.9	8.8	34.8
Vietnam	4813	2.00	9.63	4921	2.23	10.99	2.2	11.7	14.1
Other countries	342	1.85	0.63	433	1.74	0.75	26.6	-6.1	18.8
Total	113,639	2.06	234.62	122,632	2.38	291.49	7.9	15.1	24.2
AFRICA									
Egypt	348	5.30	1.84	456	5.26	2.40	31.0	-0.7	30.1
Guinea	277	1.00	0.28	411	0.89	0.37	48.4	-11.0	32.0
Ivory Coast	249	0.89	0.22	307	1.25	0.38	23.3	40.5	74.1
Madagascar	843	1.85	1.56	1026	1.77	1.82	21.7	-4.6	16.2
Nigeria	180	1.14	0.21	337	1.19	0.34	87.2	4.1	64.4
Sierra Leone	273	1.23	0.34	361	1.35	0.49	23.1	10.1	45.5
Zaire	72	0.86	0.06	275	0.77	0.21	281.9	-10.6	240.3
Other countries	958	1.02	0.98	1024	0.99	1.01	6.9	-3.5	3.2
Total	3200	1.72	5.49	4161	1.69	7.02	30.0	-1.7	27.8
SOUTH AMERICA									
Brazil	3809	1.61	6.12	4743	1.46	6.91	24.5	-9.4	12.8
Colombia	293	1.56	0.58	309	3.97	1.23	5.5	153.9	113.1
Other countries	548	2.46	1.35	686	2.86	1.96	25.2	16.1	45.4
Total	4650	1.73	8.05	5738	1.76	10.10	23.3	1.7	25.5

(Table 1 cont.)

Region or Country	1961-65			1971-75			Change (%)		
	Area (thousand ha)	Yield (t/ha)	Production (million tons)	Area (thousand ha)	Yield (t/ha)	Production (million tons)	Area	Yield	Production
NORTH AND CENTRAL AMERICA									
United States	705	4.37	3.08	902	5.07	4.57	27.9	15.9	48.2
Other countries	565	1.72	0.97	721	2.29	1.65	27.6	33.7	70.5
Total	1270	3 19	4.05	1623	3.84	6.22	27.8	20.1	53.5
OTHER									
Europe	326	4 66	1.52	395	4.57	1.81	21.2	-1.4	19.4
USSR	158	2 46	0.39	454	3.86	1.75	187.3	56.7	348.9
Oceania	35	4.57	0.16	62	5.63	0.35	77.1	23.2	119.5
World Total	123,278	2.06	254.27	135,065	2.36	318.74	9.5	14.4	25.4

[a]Data for individual countries are given only for those planting over 200,000 hectares of rice annually.

Sources: FAO, published data, except for Thailand, Indonesia, and Taiwan. For Thailand, data for 1961-65 are from Y. Gaesuwan, A. Siamwalla, and D. E. Welsch, 1974, *Thai Rice Production and Consumption Data, 1947-1970*, Department of Agricultural Economics, Kasetsart University, Bangkok. The authors of this article found the production data of the Rice Department, Ministry of Agriculture, to be underestimated, so they increased the figures by 16.29 percent. For Thailand for 1971-75, data are from Bank of Thailand *Monthly Bulletin* 17, no. 9 (September 1977). For Indonesia, data are from Central Bureau of Statistics, Indonesia. The data for dry stalk paddy were converted to paddy, using the factor of 76.47 percent. The production data for 1961-65 were adjusted by dividing by the factor of 0.86725. The Taiwan data were taken from published statistics of the U.S. Department of Agriculture.

The countries in category 3 have the potential for increasing yields, but to do so they must intensify their efforts to overcome prevailing problems of water control and poor cultivation techniques. The data in Table 1 would place Laos in category 2, but because no significant increases in yield have been achieved in that troubled country since 1972, it is included in category 3. In contrast, the Philippines has shown a steady yield increase in the same period: 1.48 t/ha in 1972, 1.74 t/ha in 1975, and an estimated 1.81 t/ha in 1976.

Africa

In Africa, Egypt and Madagascar alone accounted for 60 percent of the rice production during the 1971-75 period. The only African country to show a large increase in average yield was the Ivory Coast, and even there the yield in that 5-year period averaged only 1.25 t/ha. Zaire and Nigeria exhibited a sizable increase in area planted to rice. With their abundant river water resources for irrigation, those countries could become major African rice producers if they choose to make the effort.

South America

In South America the relatively low average yield is accounted for largely by the fact that Brazil, which plants immense expanses to rainfed upland rice with consequent low and undependable yields, contains 82 percent of the land planted to rice on the continent. The data for Colombia, on the other hand, show the progress that can be made through the widespread introduction, on irrigated land, of the modern rice varieties and the new technology. (Colombia's achievement, along with that of several other countries that have been unusually successful in developing rice production programs, is reviewed in detail in chapter 5.)

Other Areas

In the United States, Europe, and Australia (which accounts for most of Oceania's rice production), 100 percent of the rice crop is irrigated, adequate quantities of fertilizer are applied, and good farming practices prevail. The high yields (over

TABLE 2. APPARENT AVERAGE ANNUAL PER CAPITA RICE
CONSUMPTION IN SELECTED COUNTRIES, 1971-75

Country	Per capita consumption (kg)
Vietnam	239
Thailand	203
Laos	202
Burma	174
Bangladesh	161
Kampuchea	137
South Korea	136
Indonesia	121
Malaysia	113
Japan	107
Nepal	104
Philippines	89
India	73
China	72

Source: U.S.D.A. Foreign Agricultural Service, *Foreign Agriculture Circular FR 1-76*, May 1976, Washington, D.C.

5 t/ha) reflect the response to good management and to the low incidence of insects and diseases. Consequently, any further increases in rice production in those areas undoubtedly will have to come largely from an expansion in land devoted to the crop. As it is, yield estimates for 1975, 1976, and 1977 exhibit no significant increase over those for 1974.

The Soviet Union appears to have the fastest moving rice production in the world. Between 1961-65 and 1971-75 its rice-growing area increased by 187 percent, its yield level by 57 percent, and its total production by 349 percent. For 1976, the U.S. Department of Agriculture placed the Soviet rice crop at 2.2 million tons and the area sown to rice at 522,000 hectares. This brings the average yield to 4.2 t/ha. In spite of record rice production, the Soviet Union imported 250,000 tons in 1976, an indication of its high demand for rice. The tenth Soviet 5-year plan calls for further expansion, and it is estimated that by 1980 the Soviet Union will be producing 3 million tons of rice annually.

Rice as a Staple Food

Per Capita Consumption

In all Asian countries, from India and Bangladesh eastward and from Japan and South Korea southward, rice is by far the most important food crop. The apparent average annual consumption of rice for the principal rice-consuming countries of Asia is mostly over 100 kilograms a person (Table 2).

The main reason for the low consumption in such countries as the Philippines, India, and China is that crops other than rice feed a significant segment of the population. Many Filipinos, for instance, eat maize instead of rice. In India, wheat, sorghum, and maize are widely grown. The Chinese have a diversified diet. FAO figures for the food consumption of Chinese indicate that rice furnishes 698 calories per day; wheat, 267 calories; maize, 208 calories; and millet and sorghum combined, 145 calories. In the United States, by way of contrast, the average per capita consumption of rice is only 6 kilograms a year. That level is typical on the whole for most Europeans as well—with significant regional variations.

Nutritional Value

The chemical composition of the rice grain varies considerably depending upon the genetic factor of plant variety and upon such environmental influences as location and season in which grown, fertilizer treatment, degree of milling, and conditions of storage. On the average, however, a sample of milled rice grain will contain about 80 percent starch, 7.5 percent protein, 0.5 percent ash, and 12 percent water.

The starch, as in most other cereals, is a mixture of amylose and amylopectin. The proportion of these two starches has much to do with the cooking and eating qualities but does not affect nutritional value. The higher the proportion of amylose, the drier and more separated the grains are after cooking. True glutinous rices, on the other hand, are essentially 100 percent amylopectin. The grains of japonica varieties (see below) have nearly equal proportions of the two kinds of starch.

Although rice is primarily a source of carbohydrates, because of the large quantities consumed in countries that grow no other important food crop it also constitutes the principal source of protein for millions of Asians. Those who eat more than 150 kilograms of rice annually may be obtaining from 40 to 70 percent of their protein from that source alone. In fact, many nutritionists now say that adult rice-eating populations in Asia are not suffering so much from protein deficiency as from an insufficiency of total caloric intake. Children, however, from the time they are weaned until they are about 6 years of age, suffer severely from protein deficiency if fed principally on rice. The reason is that the protein requirements of growing children are high, and their stomachs cannot hold enough rice to meet their daily protein needs.

Even though the protein content of polished rice is somewhat lower than that of wheat, maize, and sorghum, the quality of the protein is considerably higher. Lysine, the most important limiting essential amino acid, constitutes about 4 percent of the protein of rice, twice the level in white flour or hulled maize. Furthermore, the percentages of threonine and methionine, two other essential amino acids, are considerably higher in rice protein than in the protein of maize, wheat, or sorghum. Thus, because of the superior quality of the protein, rice-eating peoples are able to maintain reasonably adequate protein levels in their diets. Nevertheless, rice protein does not contain enough lysine, threonine, or methionine. Consequently, for proper protein nutrition, supplementary foods such as grain legumes, meat, and fish should be part of the diets of those who consume large amounts of rice.

Like other cereals, rice is lacking in vitamins A, D, and C. It does contain small amounts of thiamine, riboflavin, and niacin. The levels of these latter vitamins are considerably higher in brown rice than in polished rice, because the B-complex vitamins are concentrated largely in the bran and germ, which are removed by milling. Home-pounding, which is still a common way poor rural people in many countries remove rice hulls and bran, leaves the grain higher in the B-complex vitamins than milled rice, however, because bran and germ are not completely removed. Polished rice that is parboiled (see chapter 3) also tends to provide larger amounts of these vitamins than are available in the nonparboiled grain.

Nutritionists recommend eating brown rice. But it is not consumed in large quantities in the tropics and subtropics, because in storage, after the rice is dehulled, the oil in the bran tends to become rancid. Furthermore, in many rice-eating societies there is even a social stigma attached to eating brown rice, probably because it is rougher and cheaper than the refined type. In addition, there is some evidence that the continual eating of brown rice can cause digestive disturbances.

The nutritional disadvantages of rice in its polished form can be overcome through enrichment. For example, beri-beri, resulting from a diet high in polished rice, can be eliminated by fortifying the rice with B-complex vitamins. As indicated earlier, similar benefits can come from eating parboiled rice.

Although low in protein, vitamins, and minerals, rice nevertheless has several distinct advantages as a food. Its carbohydrates are easily digested, which appears to explain why its marginal protein content has proved to be so nearly adequate for rice-eating peoples. Evidence exists that easily digestible carbohydrates improve protein efficiency: the net protein utilization value for rice is 63, compared with 49 for wheat and 36 for maize.

Rice is relatively nonallergenic, which means that cases of hypersensitivity to it are rare. For this reason, patients with food allergy symptoms of unknown cause are often put on a diet exclusively of rice, to which other foods are added, one by one, until the allergenic source is identified. Because of its low sodium content, a rice diet is commonly prescribed, also, for patients suffering from hypertension (high blood pressure).

An additional advantage of rice is its enduring palatability. Most consumers can eat it daily, often at consecutive meals, for a lifetime without tiring of it. So acceptable is rice that even in regions where over the ages the traditional food crops have been roots (such as cassava, yams, and sweet potatoes), any upturn in the economy soon results in a growing popular demand for rice as a staple food. In West Africa, for instance, nearly all countries are now making a concerted effort to become self-sufficient in rice production to satisfy increasing demand, and to save scarce foreign exchange by avoiding excessive imports of rice.

It seems likely that rice, in spite of deficiencies as a complete food, will continue to be consumed in large quantities by millions of people, and that its worldwide popularity will increase during the years ahead. Consequently, planners, administrators, and project agents involved in rural improvement in rice-producing countries have the twofold task of seeking opportunities to enrich national diets through crop diversification and, at the same time, of working unrelentingly to increase the yield and the total production of rice, the crop that for the foreseeable future will continue to be the major source of calories in so much of the developing world.

Types of Rice

Although there are at least 20 species of the genus *Oryza*, most cultivated rice is *Oryza sativa* L. In fact, the only other species of rice grown for food is *O. glaberrima* Steud., found solely in parts of West Africa. Its importance, however, is decreasing as it is replaced by modern varieties of *O. sativa*.

O. sativa varieties have been separated into three types: indica, japonica, and bulu. Their origin appears to be the result of selection by man in the process of domestication and selection of the wild rices under different environments, for no such natural differentiation occurs in *O. nivara*, which is considered by most authorities to be the most likely progenitor of *O. sativa*.

Although 40 years ago indica and japonica rices were thought by most rice scientists to be subspecies of *O. sativa*, they are now considered to be ecogeographic races. As hybridization between the two groups continues, probably even that distinction will disappear, because any of the identifying japonica and indica characteristics can be transferred in either direction through crossing and selection.

The Indica Type

The traditional indica rice varieties, widely grown throughout the tropics, are tall and heavy tillering (tillers are secondary stems) with long droopy light-green leaves. They exhibit little tolerance to cold temperatures and respond in grain yield only to low applications of fertilizer. However, unlike japonica varieties, many indicas possess considerable drought tolerance and resistance to insect and disease attack. In general, the grains of the indicas are medium-long to long, and the amylose content of the starch is medium to high, causing the cooked rice to be dry and fluffy and to show little disintegration.

Commonly, when indica and japonica types are crossed, there is a high degree of sterility in the F_1 generation. This sterility originally led taxonomists and breeders to decide that the japonicas and indicas belonged in separate subspecies. It is now known, however, that the sterility can be bred out of the progeny in a few generations.

Recent breeding work has produced short-statured, heavy-tillering indica rice varieties that respond to fertilizer and produce yields as high as those of japonicas. (An account of the development of modern indica rice varieties is given in chapter 2.)

The Japonica Type

The japonica varieties have greener, more erect leaves and a lower tillering capacity than the indica varieties. They are resistant to lodging and are more nitrogen responsive in their yield. Disease and insect resistance, on the average, is lower than in the indica varieties. Generally, the grain is shorter and wider. The amylose content of the starch is lower, so the cooked rice is stickier and glossier, and the grains tend to disintegrate if boiled too long.

Essentially all of the rice varieties in countries with temperate climates, such as Japan, Portugal, Spain, USSR, Italy, and France, are of the japonica type. The japonica race originated in China, and the Chinese term *keng* has been used to designate it since the first century A.D.

Both indicas and japonicas are grown in Egypt, China, Taiwan, the United States, and Australia. Recently, Korea developed new rice varieties that were japonica-indica hybrids, which, benefiting from an extensive promotion campaign, markedly increased average national yields. (This achievement is described in chapter 5 as an example of a successful national rice production program.)

The Bulu Type

In Indonesia two types of rice are widely grown: the indica type (called *cere* there) and the bulu type. The bulu rices are morphologically similar to the japonicas, but have wider and more pubescent (hairy) leaves. In addition, the grain frequently has hairlike awns (the name bulu means "bearded"; an awnless bulu is called *gundil* in Indonesian). Like the japonicas, the bulus are low tillering, have stiff straw and are relatively insensitive to photoperiod (day length).

Varieties belonging to the bulu type are grown only in Java and Bali, in the rice terraces of the Philippines, and in the mountainous areas of Madagascar, which suggests that there was communication between the peoples of those regions in ancient times. Probably bulu rices will gradually disappear from Indonesia, since the modern indicas are now spreading rather rapidly among the farmers of that country.

Oryza Glaberrima

Oryza glaberrima is grown as a food crop only in West Africa. It probably originated along the Niger River in Mali. Evidence exists that *O. glaberrima* may have been eaten there as early as 3500 years ago. Its characteristics are smooth hairless glumes, red grains, short ligules with roundish tips, high seed dormancy, and stiff upright panicles with few or no secondary branches.

Most rice specialists consider *O. glaberrima* to be inferior to *O. sativa* in yielding ability and in disease resistance. Like the indica rices, there is a tremendous range in the ecological conditions under which the *glaberrimas* are produced. Although they can be found growing under rainfed upland conditions, their most common environment is in deep-water areas. In fact, some *glaberrimas* outyield indica floating and deep-water varieties.

Recently, both the International Rice Research Institute (IRRI) in the Philippines and the upland rice research station at Bouake, Ivory Coast (staffed and partly funded by the Institut des Recherches Agronomiques Tropicales, Paris), have assembled collections of over 1000 varieties of *O. glaberrima.* Now that it is possible for plant breeders to grow and observe large numbers of *glaberrima* varieties together, there is renewed interest in hybridizing *glaberrimas* and indicas. The considerable variation among the *glaberrimas* with respect to plant type, tillering capacity, stem thickness, and insect and disease resistance has become appreciated only recently. Because of the wide genetic gap between the two kinds of rice, problems of sterility arise when crosses are made. But by making many crosses and selecting the few that are successful, the sterility can be bred out of the progeny in a few generations.

Quality Preferences among National Groups

Rice is unique among the cereal grains in that the entire polished grain is eaten. Consumer preferences regarding, and prejudices about, the cooking and eating quality of rice vary considerably from region to region. In addition to environmental suitability, consumer preferences have much to do with which rice varieties are grown in a given area.

The cooking and eating qualities of rice are determined largely by the starch composition of the grain. The two starch fractions, amylose and amylopectin, are distinguished by the arrangement of their glucose units. In amylose, the glucose units are arranged linearly, while in amylopectin they are branched. The higher the proportion of amylose (and consequently the lower the proportion of amylopectin), the greater the tendency of the rice to cook dry and fluffy, and the greater the resistance of the grain to disintegration even after prolonged cooking. Low amylose rices tend to be stickier, glossier, more tender, and more likely to disintegrate if overcooked. Other factors—particularly variations in the tendency of cooked rice to harden when it cools—influence the cooking and eating qualities of rice, but by far the most important is the ratio of amylose to amylopectin. In "glutinous" or "waxy" rices, the starch is essentially 100 percent amylopectin. The amylose content of nonwaxy milled rice is classified as low (below 20 percent), medium (20 to 25 percent), and high (above 25 percent).

In general, rice eaters in India, Pakistan, Bangladesh, Malaysia, Sri Lanka, southern China, and most Latin American countries consume high amylose rice. However, they prefer those varieties that have a soft gel consistency, meaning that the cooked grains do not harden excessively when they cool.

In the Philippines and Indonesia, there is a definite preference for medium amylose varieties that are not as hard and dry-cooking as those of India, but not as sticky and glossy as those favored in Japan. In spite of this preference, quite a few high amylose varieties are grown in both the Philippines and Indonesia.

The japonica varieties grown in Japan, in central and northern China, and in other countries mentioned earlier, are all low amylose in character, containing 12 to 18 percent of that component.

The people of Laos and of most of north and northeast Thailand use the true glutinous rice for their staple food. Most countries, in fact, grow some glutinous rice for special uses in cakes, confections, and similar dishes.

The *basmati* rice varieties of northern India and of Pakistan have long, slender grains that elongate greatly when cooked (to about double their raw size) and are strongly aromatic. These varieties command an excellent price on the Mideastern and European markets and provide valuable export earnings for the countries producing them. In addition, they have a market among the more affluent local families. Scientists have not yet discovered the cause for the extraordinary elongation of the grains during cooking. It is related neither to amylose content nor (consistently) to any other characteristic that has been measured.

In addition to the inherent cooking and eating qualities of rice varieties, a process called parboiling (see chapter 3) produces distinctive characteristics that are preferred by many consumers in India, Sri Lanka, and Nigeria, for example. In many other countries, parboiled rice is not acceptable to the mass of consumers.

Rice as a Unique Food Crop

Vast areas of flat, low-lying tropical soils in Asia are flooded annually during the rainy monsoon season. With the exception of taro, rice is the only major food crop that can be grown in standing water. Wheat, maize, sorghum, yams, sweet potatoes, white potatoes, and cassava, to mention some of the world's most important food crops, would die in the continuously submerged soil in which rice not only survives but thrives. Rice is uniquely adapted for growth in submerged soils because it possesses tubes in its leaves, stems, and roots that permit air to move from the leaves to the root surfaces, thus supplying the submerged roots with sufficient oxygen for normal respiration and nutrient absorption. This semiaquatic nature of the plant allows it to be grown in the many great river basins and deltas of tropical and subtropical Asia, where it provides the principal food for the multitudes who dwell there. Were it not for rice, those areas would undoubtedly be unable to support even one-quarter of their present populations.

Although the highest yields of rice are obtained with controlled water depths of less than 15 centimeters, rice can be grown under upland conditions with no flooding, and some varieties can tolerate water depths up to several meters.

Rice is one of the few crop plants that can be grown on the same land year after year without serious problems developing.

The high silicon content of rice hulls and, to a lesser degree, of the leaves and stems enhances the plant's ability to resist the attack of certain insects and diseases.

Rice grows under a wide range of soil acidity and alkalinity. To a large extent its "tolerance" to variations in the soil pH stems from the ability of rice to grow in submerged soil and the fact that, under water, the pH of acid soils increases and that of alkaline soils decreases. (The chemistry of flooded soil is discussed briefly in chapter 2.)

Through natural selection by farmers over the ages and by rice scientists more recently, some varieties of rice are quite tolerant to soil salinity and to such other adverse soil conditions as phosphorus and zinc deficiency and iron toxicity.

Rice breeders today are working with soil scientists to breed rice for tolerance to adverse soil conditions.

Although cultivated rice appears to have originated in the tropics, varieties have been developed in the past half-century that produce high yields in cool areas, such as Hokkaido, Japan, and northern China. Some cold-tolerant varieties can be grown at elevations of 2400 meters in the tropics. However, rice is not an important crop under such conditions.

No other leading food crop is so adaptable to such a broad range of climatic and soil conditions.

Kinds of Rice Culture

As already stated, rice is produced under lowland and upland conditions as well as in water depths up to several meters. The following brief descriptions of the growing of rice under four different conditions are obviously explanatory in purpose and are not at all intended as instructions in rice management.

Irrigated Lowland Rice

Rice grown in bunded (diked) fields in which irrigation systems fed from wells or rivers control the depth of water is called irrigated lowland rice. In Asia, most rice grown under such conditions is transplanted, fertilizer is usually applied, and yields, on the average, are higher than those under any other system of rice growing. Nearly all rice in the high-yield countries such as Japan, South Korea, and China is grown under irrigation and with provision for drainage where needed, thus ensuring a controlled water supply.

In Asia, the recommended procedure for managing irrigated rice is to flood the paddies, to plow and harrow (and level the land if necessary), to apply fertilizer before the last harrowing, and then to transplant the rice seedlings, which should not be over 21 days old. In the tropics controlled year-round irrigation facilities permit the growing of two or more rice crops a year, or one or two of rice as well as other crops (that is, multiple or relay cropping).

Rainfed Paddy

The most common system of growing rice in South and

Southeast Asia is called rainfed paddy. The paddies are bunded (diked), and after the monsoon rains come and water accumulates on the soil surface, the land is plowed and harrowed. The subsequent management of the crop is similar to that for irrigated rice.

If rainfall is adequate and evenly distributed, rice yields can be as high on rainfed paddies as on irrigated fields during the monsoon season. However, too often poor rainfall depresses yields. For that reason, farmers and governments have installed supplemental irrigation facilities for many areas of rainfed paddy rice. But in other areas rainfall may be excessive and drainage facilities may be needed. Rice yields on rainfed paddy fields also tend to be lower, because farmers are less likely to spend money on such inputs as fertilizers and pesticides when there is uncertainty about the amount and distribution of rainfall.

In some tropical regions it is possible to direct-seed the first crop of rice as soon as the soil is wet enough to be plowed and harrowed, but before there is enough water in the paddy to flood it. As this first crop is approaching harvest, a seedling nursery is planted; and after the harvest of the direct-seeded crop, a second crop is transplanted. This method requires the use of herbicides on the direct-seeded rice and also, of course, the use of early maturing varieties. Although it has proven successful in some areas, more research and testing are needed before the method can be recommended generally to farmers.

Upland Rice

Rice grown in fields like any other cereal is called upland rice. Under this system of rice growing, the land is tilled before the rainy season arrives, and the rice seed is broadcast. Little fertilizer is applied. There is no bunding, so water cannot be held on the land surface. Weed control is difficult, and yields on farmers' fields average less than 2 t/ha. In bad years they may average less than a ton.

By seeding in rows, cultivating the rice in the early stages for better weed control (or by using herbicides), and by using fertilizers and pesticides, yield levels can be raised substantially. However, because of erratic rainfall patterns, poor weed control, low fertilizer use, and high disease incidence, yields normally remain low. Many authorities recommend that upland areas that cannot be economically bunded, or that have sandy soil types, be converted to the growing of crops such as maize, sorghum, soybeans, or sweet potatoes that have much more drought tolerance than rice. Rice breeders are attempting to develop upland varieties with higher levels of drought tolerance or a higher base yield than are found in present varieties. It is too soon to tell how successful this effort will be.

Deep-water or Floating Rice

Rice grown in areas where water depths reach 1 to 5 meters is called deep-water or floating rice. Widespread sections of Bangladesh, Thailand, India, Vietnam, and Indonesia flood to such depths every year. West Africa has extensive areas of deep-flooded land on which no other crop but rice can be grown.

Varieties that are planted for deep-water conditions have special genetic characteristics that enable them to survive and grow. Their internodes are able to elongate (as much as 10 centimeters per day) as flood waters rise. The varieties have the ability to produce adventitious roots at the upper nodes. And their photoperiod sensitivity keeps them from reaching maturity before the flood waters recede.

The usual system of culture is to prepare the soil when it is dry or just after the first rains. The rice seed is then scattered by broadcasting. If the rains fall regularly and the rice plants become well established, reasonably good stands are obtained. If drought occurs or if the floodwaters are late, poor stands may result from drought or from weed competition. Normally, deep-water rice is not fertilized.

The term "floating rice" is used, because at times the plants may be uprooted by strong winds and heavy currents and actually float away. However, while moving they continue to draw nutrients from the flood waters through their adventitious roots; and, as the flood subsides, the plants become reestablished in the soil and may even put out a new set of tillers.

Under farm conditions, the yields of deep-water rice, like those of upland rice, seldom exceed 2 t/ha. Nevertheless, no food crop other than deep-water rice can be grown in such flooded areas. (Chapter 6 discusses the potential for developing new rice varieties that have greater tolerance to deep-water conditions.)

Future Supplies of Rice

Many studies have been conducted by the FAO, by govern ments, by privately supported organizations, by university teams, and by individual social scientists in an effort to predict future production of food, including rice. Generally, such studies have made estimates extending either to 1985 or to the year 2000. Because of the uncertainty of future population growth rates and of economic development, some surveys have made as many as four predictions of the future supply and demand for food. Rather than an effort to describe here the numerous detailed studies on the subject, there follows a description of several factors influencing such forecasts of the human food situation, plus a kind of "average" conclusion that could be made.

Naturally, predictions must be based largely on past performances and, more particularly, on recent trends. Because nearly 90 percent of the world's rice is produced and consumed in low-income countries, the most reliable evidence for future forecasts can be found by examining recent trends in rice production and in population growth in the less developed countries.

Figure 1 shows that rice production has gained only slightly on population growth since 1950-52. Thus the output per capita has remained rather constant. Of course as economic advance occurs, there is a tendency (as in Japan) for the consumption of rice to decrease and for that of higher protein food to increase. A comparable graph for the developed countries would show that food production, including that of rice, has increased more rapidly than has population.

In simple terms, the future per capita production of rice will depend on the land area devoted to the crop, the intensity of multiple cropping, the average yields obtained, and the rates of population growth. Per capita demand would also affect the outcome. However, the various studies of the demand for rice

Index no. (1949/50 - 1951/52 = 100)

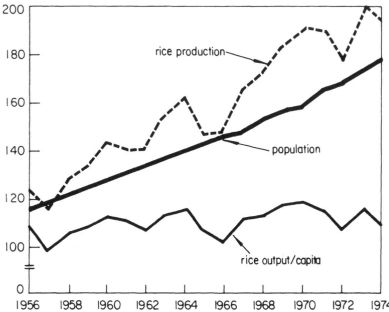

Figure 1.   Production, population, and per capita output of rice in the less developed countries, 1956-74.

have revealed no indication that per capita rice consumption will decline. In fact, a modest increase is forecast through the year 2000.

During the first half of the twentieth century, many Asian countries exported rice. Today, however, although per capita consumption has shown little change, only a few countries in Asia—principally Thailand, Burma and Pakistan—have exportable surpluses. Thailand alone has been able to maintain its rice exports at a fairly high level (about a million tons a year). As its population continues to grow, however, the domestic demand for rice likewise will increase, and Thailand ultimately may be without an exportable rice surplus. China is a large exporter of rice, but it simultaneously purchases even larger amounts of wheat, because wheat prices on the world market are lower than those for rice. Consequently, it can purchase more calories in wheat than it loses in the rice it sells. Most Asian countries today are struggling to attain self-sufficiency in rice. Although recently the Philippines and

India have achieved rice self-sufficiency, it is doubtful that this represents more than temporary surpluses.

Until the 1950s, yield levels of rice were rather constant and increases in production came from gains in the land area planted to the crop. Inevitably, the dependence on higher yield rather than on expansion in area will increase as time goes on. In fact, most of the recent increases in land area can be attributed to double cropping—that is, the growing of two crops of rice a year on the same land on which only one crop was grown before. (The FAO counts such land twice, so that yields refer to a single crop, though two crops were grown in a single year on the same land.)

It is clear that future production increases in Asia will come largely from improving yields on land already growing rice and from expansion of double cropping of the grain. Fortunately, current yields in many Asian countries are so low that there is considerable room for improvement. Countries that are producing less than 2 t/ha of paddy rice conceivably can double their production by the time their populations double. On the other hand, Japan and South Korea, with 1976 yield estimates well over 5.5 t/ha, cannot be expected to increase their yields much beyond those high levels. Even if they succeed in doing so, their harvests are likely to satisfy only their domestic demands.

China, with an average yield of over 3.2 t/ha obtained on 34 million hectares, had little increase in yield between 1974 and 1976 according to FAO statistics. However, rice scientists who have recently visited China believe that, at full operating capacity, the country's new fertilizer plants could increase average rice yields by about 1.2 t/ha. Japan went through a similar static period in the 1920s and 1930s, and its rice yields moved upward again when chemical fertilizers began to be used in addition to organic manures.

The only Asian countries to have brought large new areas into rice production recently are Thailand and Indonesia. According to the FAO, during 1974-76 Thailand added over 1 million hectares of land to its rice-growing areas, most of which was rainfed paddy. Indonesia is developing rice lands on its less populated islands, since Java, Madura, and Bali are greatly overcrowded and essentially all of the suitable rice land on those islands is already planted to the crop. Even Thailand is unlikely to bring much new rice land into cultivation in the years ahead. Instead it will continue to develop its irrigation systems and to depend upon double cropping and the use of modern technology to increase its rice production. This, of course, will be true for all Asian rice-growing countries except the few that already irrigate all of their rice crop.

Many analysts predict that in future years Asia will reduce its upland rice area, substituting more drought-resistant crops for rice on those lands. They foresee, also, yields on lowland areas—both irrigated and rainfed—rising considerably as better varieties are planted and as improved methods of weed control, fertilizer use, and pest control are more widely practiced.

Rice production undoubtedly will increase in Latin America and in Africa, with the early gains coming mainly from expansion in land area devoted to rice. In the long run, however, as new irrigation systems are built, yields also will increase. Colombia has already achieved remarkable yield improvements on its irrigated rice land. In spite of these prospective increases in production in Latin America and Africa, the evidence indicates that for well into the twenty-first century the large preponderance of the world's rice will continue to be produced in Asia, where so much of the rice-eating population will still be found.

To sum up, it can be said that all studies have predicted a food deficit in the developing countries as a whole by the year 2000 unless huge investments are made in agricultural improvement and unless population growth rates decrease considerably. The estimates for rice production by the end of this century are somewhat more optimistic, indicating that with expanded irrigation systems, new varieties, increased fertilizer use, and the necessary infrastructure, rice supply may continue to follow the traditional trend and just keep pace with population growth. This is not a favorable outlook, however, for it means that present low-nutrition standards are unlikely to be improved.

The United Nations estimates that global population will

reach 6000 million by the year 2000, as compared with just over 4000 million in 1976. If their predictions materialize, India will have a population of over 1000 million by that time.

In looking ahead, it is useful to examine the practical rice production potential of the Asian countries that are now growing much of the world's rice. For reasons already mentioned, it seems likely that South Korea, Japan, and China will have difficulty in keeping up with their domestic needs after a decade or two. The other Asian countries, however, can produce considerably more rice, because their current yields are so low.

The best effort to assess future production potentials has been made by economists at the International Rice Research Institute in the Philippines. In their study, the actual yields of trials on farmers' fields were used as a gauge of what could be accomplished if modern varieties and improved technology were employed. Such yield levels, of course, are lower than those obtained on experimental fields where environmental conditions are highly controlled.

Because levels of yield vary considerably under the different kinds of rice culture, the IRRI investigators separated the land area of each of 11 rice-growing countries into four categories for which they then estimated the potential yield on farmers' fields: (1) dry-season irrigated (100 percent irrigated); (2) wet-season irrigated (supplemental irrigation in the wet season only); (3) rainfed paddy (bunded but with no supplemental irrigation); and (4) deep-water and upland rice (these two classes were thrown together only because the yield potentials appear to be similar).

From the results of the tests on farmers' fields and the knowledge of the approximate area of each class of rice land in each of the 11 countries, it was possible to obtain a weighted average of the maximum yield that appears to be possible under farm conditions. The researchers' estimates of area in each of the four categories and their weighted averages for each of the 11 countries are shown in Table 3.

For all countries listed the potential yield on dry-season irrigated land is higher than it is under any other condition. This points out the combined benefits of irrigation and solar

TABLE 3. ESTIMATED MAXIMUM FARM YIELDS FOR 11 ASIAN COUNTRIES, AND THE AREA OF MAJOR TYPES[a] OF RICE LAND OF VARYING YIELD POTENTIAL

Country	Area[b] (thousand hectares)				Maximum yield[c] (t/ha)				Weighted average (t/ha)
	DS	WS	UD	RF	DS	WS	UD	RF	
Philippines	480	1430	380	1200	5.9	4.6	2.0	3.5	4.1
India	1890	13,120	3770	18,880	6.8	5.4	2.0	4.0	4.4
Indonesia	1610	2370	1870	2630	5.9	4.8	2.0	3.6	4.0
Thailand	140	630	630	5630	4.4	3.7	2.0	2.5	2.7
Bangladesh	980	590	4390	3810	6.6	4.9	2.0	3.7	3.3
Vietnam	140	270	680	1630	5.8	4.1	2.0	3.1	3.1
Sri Lanka	150	220	10	220	5.7	5.3	2.0	4.0	4.8
Burma	50	800	100	4040	6.0	4.8	2.0	3.6	3.8
Pakistan	1520	0	0	0	6.0	—	—	—	6.0
Nepal	0	190	110	910	—	4.8	2.0	3.6	3.7
Malaysia	270	270	20	150	6.0	4.8	2.0	3.6	4.5

[a]DS = dry-season irrigated; WS = supplemental irrigation, wet season only; UD = total area upland rice (nonbunded fields) and deep-water rice; RF = rainfed paddy, bunded but without irrigation.

[b]Based on IRRI estimates of proportions in each of the four types, and the FAO figures for total area planted to rice (1970-74 averages).

[c]Average maximum yields obtained on trials in farmers' fields in the Philippines, India, Indonesia, Thailand, Bangladesh, Vietnam, and Sri Lanka. For the other four countries, maximum possible yields were taken as the average of the other seven countries.

energy in obtaining high yields of rice. Supplemental irrigation in the wet season gave the next highest calculated maximum yield, ranging from 3.7 to 5.4 t/ha. The estimate for maximum on-farm yields under rainfed paddy conditions, with no supplemental irrigation, gave values from 2.5 to 4.0 t/ha. The IRRI economists assumed a 2.0 t/ha yield for upland and deep-water rice. The figure was based on a number of trials under those conditions, but because yields are so variable from season to season, depending on the amount of rain and on its distribution pattern, the scientists decided to assume a constant for each country.

The weighted averages shown in Table 3 represent the best estimate of the mean maximum yield obtainable under farm conditions in each country, assuming that the areas and proportions of land in each of the four categories do not change. These maximum calculated national yields vary from 2.7 t/ha for Thailand to 6.0 t/ha for Pakistan, which grows only irrigated rice.

So that officials in the 11 Asian countries listed in Table 3 might gain a practical view of the amount of rice that their countries could produce, Table 4 was constructed. The data indicate that it is possible to double rice production on the farms of the 11 Asian countries, even with no more land under irrigation than at present. These optimistic estimates of maximum production, however, assume that all farmers will follow the recommended practices faithfully and fully—a condition that experience has shown is not likely to come about. China has made an herculean effort to produce more rice, yet its average yields are still less than 3.5 t/ha. A more likely prediction of rice yields in those 11 countries for the year 2000 would seem to be about 3 t/ha. This means, then, that rice production would more or less keep abreast of population growth in the intervening time. Even that modified figure probably will not be attained without major efforts on the part of the leaders in all rice-producing countries to increase the amount of irrigated rice land and to make available to farmers the inputs and the necessary infrastructure. (The requirements for launching and maintaining a successful rice production program are discussed in chapters 7 and 8.)

TABLE 4. AREA, YIELD, AND PRODUCTION OF PADDY RICE IN 11 ASIAN COUNTRIES IN 1976, THE WEIGHTED AVERAGE OF THE CALCULATED ON-FARM MAXIMUM YIELDS (FROM TABLE 3) AND THE ESTIMATED AMOUNT OF RICE THAT COULD BE PRODUCED IN EACH COUNTRY

Country	1976			Calculated	
	Area (million ha)	Yield (t/ha)	Production (million tons)	Average maximum yield (t/ha)	Maximum production (million tons)
Philippines	3.62	1.8	6.7	4.1	14.8
India	36.00	1.8	69.0	4.4	158.4
Indonesia	8.80	2.6	23.0	4.0	35.2
Thailand	8.20	1.8	14.5	2.7	22.1
Bangladesh	9.90	1.8	18.2	3.3	32.7
Vietnam	2.30	2.0	4.5	3.1	7.1
Sri Lanka	0.52	2.0	1.0	4.8	2.5
Burma	5.13	1.9	9.5	3.8	19.5
Pakistan	1.70	2.2	3.8	6.0	10.2
Nepal	1.27	2.1	2.6	3.7	4.7
Malaysia	0.58	3.2	1.8	4.5	2.6
Total or avg.	78.02	2.0	154.6	4.0	309.8

Finally, it must be remembered that the foregoing predictions do not extend beyond the year 2000, whereas mankind's struggle for survival in the face of its increasing numbers will continue indefinitely. Eventually, the entire system under which the world's peoples live—economic, social, and technological—must be brought into a state of equilibrium. The United Nations predicts that the earth's population might not become stabilized before the year 2100 or even 2150, by which time it will be between 10,000 million and 16,000 million. It is sobering indeed to consider the enormous resources of food, energy, and water required to support a population of that size, quite apart from the concern over the social, economic, and political problems engendered by such overcrowding of the human species.

2
The Modern Rice Plant and the New Technology: Greater Potentials for Rice Production in the Tropics

Since 1960 more progress has been made in increasing the yield potential of the tropical rice plant than had occurred in the first 50 years of this century. As rice breeders were developing the new higher yielding varieties, other scientists were gaining more knowledge of the physiology and the chemistry of rice. Some specialists were improving methods of insect and disease control, of water management and of fertilizer use; others developed small, relatively inexpensive agricultural machines and equipment for use in the low-income countries. Soil scientists revealed the changes taking place in the flooded rice soils under tropical conditions; agricultural economists studied the factors affecting farmer adoption of the new technology and the relationship between such acceptance and farm income.

This chapter presents some of the more important recent advances in rice science in the tropics, to provide, especially for persons who are not rice scientists, a concise account of the modern rice technology that has so decidedly widened the scope for increasing rice production in the tropics.

Although many of the studies discussed in the following sections were conducted by the International Rice Research Institute (IRRI) in the Philippines, the rapid progress they represent could not have been made without benefit of the research conducted earlier in other parts of Asia, such as Japan and Taiwan. Furthermore, the work at IRRI was carried on by an international (largely Asian) team. Such scientists brought to IRRI a knowledge of and an experience with rice that were invaluable in getting the research program off to a strong start. In addition, the continual involvement of scientists of other countries in IRRI-based symposia and in cooperative research and testing programs in their homelands contributed, and is still contributing, to the success of IRRI.

The Modern Tropical Rice Plant

The traditional tropical rice plant, an indica type, is tall (usually 160 to 200 centimeters) with long drooping leaves. It has been bred and selected for dependable yields under low management levels, is tolerant to variations in water level in the paddies and competes reasonably well with weeds. It endures low soil fertility conditions and is fairly resistant to insect and disease attack. But the yield potential of the traditional varieties is low. Yields on farmers' fields seldom exceed 2.5 t/ha and often fall considerably short of 2 t/ha.

As the availability of new rice land in Asia dwindled in the 1950s, efforts were made to use modern inputs, particularly fertilizer, on the traditional rice varieties in an effort to raise their yields. The results were disappointing. When nitrogen was applied to those tall leafy indica rice varieties, they became even taller and leafier, so that by flowering time or soon after they lodged (toppled over), thereby decreasing yields markedly.

Scientists soon learned that the earlier in its life that the rice plant lodged, the greater the reduction in yield. In 1955, for example, most tropical rice research stations recommended that no more than 30 kg/ha of nitrogen be applied to rice, because larger amounts either did not increase yields or actually depressed them. The result was that even on experimental fields, with near perfect water control and with other cultural practices at ideal levels, essentially no yields of more than 2.7 t/ha were reported; indeed, most were lower. At the same time, Japanese agronomists (with japonica varieties) were obtaining yields well over twice that quantity.

The dilemma thus facing the tropical Asian farmer was as follows: if he used fertilizer, particularly in the rainy season, either he got little in return for it or his yields decreased because of lodging; if he applied no fertilizer, yields remained low, because the rice was undernourished.

Although rice breeding took place in the 1920s and 1930s in Japan, Taiwan, India, the Philippines, and elsewhere, it was not until after World War II that the shortage of food and the ominous population increase caused major attention to be focused on improving the yield potential of rice in Asia, where it was by far the most important food crop.

One of the important programs of the 1950s was the indica-japonica hybridization project of the Rice Breeding Working Party of the International Rice Commission of the FAO. The Asian countries participating in the project sent their most promising varieties for crossing with japonicas to the Central Rice Research Institute in Cuttack, India. From the crosses made there, F_2 generation seed was sent to the cooperating countries for testing and selection. Although for several reasons this project was less successful than it would have been a decade later, several superior varieties came from it, particularly ADT 27 in Tamil Nadu state, India, and Malinja and Mashuri in Malaysia. Even today Mashuri is preferred by many farmers in both Malaysia and parts of India.

National programs were also producing higher yielding varieties independent of the indica-japonica program of the FAO. Typical examples of improved indica rice varieties developed during the mid-1950s are BPI-76 in the Philippines and H-4 and H-5 in Sri Lanka (then called Ceylon). In general these varieties were earlier maturing, more disease resistant, less photoperiod sensitive, and somewhat shorter and hence more nitrogen responsive than the typical indica varieties grown in tropical Asia. Nevertheless, they could not stand high levels of nitrogen fertilizer without lodging.

When the International Rice Research Institute started its research program in 1962, its scientists were aware of the fact that in Japan, Taiwan, and elsewhere, short, stiff-strawed rice varieties had been developed that were fertilizer responsive and that had a much higher yield potential than did the traditional tall, leafy tropical varieties or even the improved varieties developed in South and Southeast Asia during the 1950s. Also, it was known that if the Japanese varieties were planted in the tropics, they flowered too early, were too low tillering for the less exact rice spacing that was practiced by the tropical rice farmer, and were highly susceptible to insect and disease attack.

Scientists in Taiwan had developed japonica varieties that were quite well adapted to the higher temperatures and shorter day lengths of the tropics, but those varieties proved to be deficient in insect and disease resistance, and they retained the low-tillering characteristic of the typical japonica plant. Scientists in Taiwan had also, however, released a short-strawed indica variety, Taichung Native 1, in 1956. From about 1960 onward it became popular with farmers there mainly because it gave yields of 6 to 8 t/ha under proper management.

Taichung Native 1 originated from a cross between Dee-geo-woo-gen, a short heavy-tillering variety, and Tsai-yuan-chung, a tall, disease- and drought-resistant variety. IRRI plant breeders were able to get seed from Taiwan of not only Taichung Native 1 but also of its dwarf parent, Dee-geo-woo-gen. In addition, they brought in another Taiwanese dwarf variety, I-geo-tse. In their effort to develop improved short-statured tropical varieties as soon as possible, IRRI scientists used these three dwarf varieties from Taiwan in many of the crosses made during the institute's first year of research activity. The other parents were tall indica varieties that were popular in the Asian tropics.

The most successful of these crosses was between Peta, a tall Indonesian variety then being grown rather extensively in the Philippines, and Dee-geo-woo-gen. Of the many selections from this cross that were tested, one, designated IR8-288-3, proved to be outstanding. After being widely tried throughout tropical and subtropical Asia in 1965, it became IRRI's first named variety in 1966: it was called IR8. In tests not only at IRRI but at other rice experiment stations in Asia, IR8 yielded from 4 to over 8 t/ha, which was more than twice the normal yield of the tall lodging-susceptible rice varieties then being grown by Asian rice farmers.

The meaningful difference between the two contrasting kinds of rice—the modern and the traditional—is one of plant type, sometimes referred to as plant architecture. The IR8 plant type is even today considered to be ideal (no variety has exceeded IR8 in true yield potential). When grown under good management, IR8 has the following characteristics:

- Short stature—the height ranging between 90 and 100 centimeters
- Short, thick, sturdy stems imparting resistance to lodging at high nitrogen levels and in strong winds and heavy rains
- Rather short and erect leaves of medium width, which allow the penetration of sunlight, consequently improving photosynthetic efficiency
- High tillering capacity, which aids in producing more panicles per unit area of land, allowing the crop stand to compensate to a degree for missing plants or, in direct-seeded rice, for thinly sown areas
- A high grain to straw ratio, or harvest index. Usually the weights of grain and straw are about equal (whereas in the traditional tall varieties no more than one-third of the total dry matter is grain)
- Photoperiod insensitivity. Although not a feature of plant type, this character is nevertheless significant, for it means that the variety has about the same growth duration in the tropics regardless of the month in which it is planted

Although all these characteristics are beneficial when rice is grown under controlled water conditions and at high soil fertility levels, the most important features are short stature and sturdy stems.

Geneticists studied the mode of inheritance of short stature and found that it is controlled by a single recessive gene. This means that when a tall and short variety are crossed, the first-generation (F_1) plants are all tall. In the second generation (F_2), three-fourths of the plants are tall and one-fourth are short. The short plants will produce only short progeny from then on. Plant physiologists discovered that this single recessive gene for shortness was the one that lowers the content of giberellic acid, the substance that promotes cell enlargement.

IR8 has its defects. Its grain is too chalky and coarse to command top prices in most Asian markets. Moreover, the

IR8 growing next to its parents, Peta (tall) and Dee-geo-woo-gen (short). (Source: IRRI)

amylose content of the grain is too high and it has a hard gel consistency that causes the cooked rice to harden when it cools. Furthermore, IR8 is not sufficiently resistant to several of the major rice insects and diseases. Nevertheless, it set a standard for improved rice plant type and established new records for yield in the tropics and subtropics. Not until the new short-statured indica rice varieties were created did scientists and farmers realize that yields of 5 to 6 t/ha were possible in the tropical rainy season and that, under highly favorable conditions, 7 to 9 t/ha could be produced in the dry season. IR8's superiority to Taichung Native 1, a progeny of a Dee-geo-woo-gen cross made a decade earlier in Taiwan, lies primarily in its better disease resistance and stiffer stems.

Both the advantages and the disadvantages of IR8 and of Taichung Native 1 stimulated rice breeders at IRRI and at national rice experiment stations in the tropics and subtropics to cross their tall local varieties with the modern short ones. They sought to develop high yielding, fertilizer-responsive varieties that had greater resistance to insect and disease attack and superior eating and cooking qualities as well. During the 1970s, many improved varieties have been developed. The work

Variety	Growth duration (days)	Diseases				Insects					
		Blast	Bacterial blight	Grassy stunt	Tungro	Brown planthopper			Green leafhopper	Stem borer	Gall midge*
						Bt 1	Bt 2	Bt 3			
IR8	130	MR	S	S	S	S	S	S	R	MS	S
IR5	140	S	S	S	S	S	S	S	R	S	S
IR20	130	MR	R	S	R	S	S	S	R	MS	S
IR22	125	S	R	S	S	S	S	S	S	S	S
IR24	125	S	S	S	MR	S	S	S	R	S	S
IR26	130	MR	R	R	R	R	R	R	R	MR	S
IR28	105	R	R	R	R	R	R	R	R	MR	S
IR29	115	R	R	R	R	R	R	R	R	MR	S
IR30	108	MS	R	R	R	R	R	R	R	MR	S
IR32	140	MR	R	R	R	R	R	MR	R	MR	R
IR34	120	R	R	R	R	R	R	R	R	MR	S
IR36	110	MR	R	R	R	R	R	S	R	MR	R
IR38	125	R	R	R	R	R	R	S	R	MR	R
IR40	115	R	R	R	R	R	R	S	R	MR	R
IR42	140	MR	R	R	R	R	R	S	R	MR	R

R = resistant MR = moderately resistant MS = moderately susceptible S = susceptible Bt 1 = biotype 1 Bt 2 = biotype 2 Bt 3 = biotype 3
* Rated in some locations in India.

Figure 2. Resistance ratings of IRRI rice varieties in the Philippines. (Source: IRRI)

at IRRI is a good example of the progress that has been made. The most recently developed varieties have proved to be resistant to eight major insects and diseases (Figure 2), whereas IR8 was resistant to only two.

By 1977, 25 to 30 percent of the rice land of South and Southeast Asia had been planted to modern rice varieties that had a high yield potential. Unfortunately, the true yield capacity of such varieties is seldom expressed on farmers' fields. In the main, the cultural practices on most farms have not been of a sufficiently high level to allow the varieties to yield well. Too often fields lack water, fertilizer applications are too low, and few pesticides are used. But farmers also need varieties that are better adapted to diverse growing conditions. To achieve this goal, plant breeders are placing more emphasis on developing varieties that have greater resistance to drought and that are intermediate in height, so that they are more tolerant of variations in water depth and can compete more successfully with weeds.

Although farm yields could be considerably higher than they are, it should be understood that obtaining the absolute maximum yield potential of any variety is not profitable. Studies have shown that the relatively small yield increases that occur in response to the application of very high levels of inputs generally are not economical.

Response of Modern Varieties to Fertilizer

The impact of the improved plant type is most evident when substantial amounts of nitrogen fertilizer are applied to the rice crop. At high nitrogen levels the modern varieties tiller heavily, produce more grains per unit area of land, and, of course, remain standing until harvest. The tall, lodging-susceptible traditional varieties, on the contrary, seldom respond to more than 30 or 40 kg/ha of nitrogen, and on fertile soils may show no response whatsoever to applied nitrogen.

Fertilizer trials comparing modern and traditional varieties have been conducted in essentially all rice-producing countries, with similar results. The average yield data obtained over a 5-year period at four locations in the Philippines, in both dry and wet seasons, are shown in Figure 3. IR8 and IR20 are modern varieties, and Peta is a traditional variety.

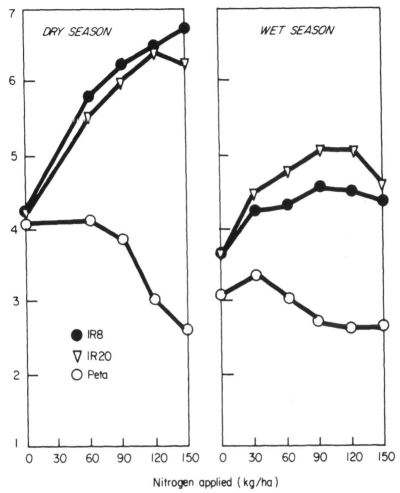

Grain yield (t/ha)

Figure 3. Effect of levels of nitrogen on grain yield of IR8, IR20, and Peta. Average of four locations (IRRI, Maligaya, Bicol, and Visayas) in the Philippines, 1968-73. (Source: S. K. DeDatta et al., 1974, *Proc. FAI-FAO Seminar on Optimising Agricultural Production under Limited Availability of Fertilizers*, New Delhi)

The modern varieties respond like the traditional varieties to phosphorus and potassium, when the soil is deficient in either of those elements. However, excess applications of phosphorus and potassium, unlike those of nitrogen, do not cause drastic yield reductions even in the traditional varieties. In other words, the modern and traditional varieties respond more similarly to phosphorus and potassium than they do to nitrogen.

In Laguna province in the Philippines in 1967, modern varieties were spreading from farm to farm. The land on the right was still being planted to Intan, a traditional variety, while the farmer controlling the land on the left had shifted to IR8. By the following season, both farmers were growing a modern fertilizer-responsive variety, for it had become evident that on fertile rice soil yields could be doubled simply by changing the variety. (Source: IRRI)

Furthermore, there is evidence that when phosphorus—and, to a lesser degree, potassium—is applied and nitrogen is not, the result is an increase in the amount of soil nitrogen available to the rice plant. The most likely explanation for this phenomenon is that with the application of phosphorus and potassium, the soil microorganisms become more active, some fixing atmospheric nitrogen and others decomposing the organic matter and thus mineralizing organic nitrogen to the inorganic form.

Soils that are alkaline in reaction or that are high in organic matter may be deficient in zinc. When field trials show a response to that element, the deficiency can be remedied easily by dipping the rice seedlings at planting time in a 2 percent suspension of zinc oxide.

Although silicon is not considered essential for plant growth, scientists have found that if the rice straw contains less than 11 percent silica (SiO_2), yield levels sometimes will rise after that compound is applied.

The efficiency of nitrogen fertilizer use for rice is low under average management conditions. Research results indicate that as much as 50 percent of the nitrogen may never reach the crop. Recent increases in the price of fertilizer have caused renewed interest in ways to heighten the efficiency of nitrogen when applied as a chemical fertilizer.

When ammonium sulfate or urea is applied to rice soils, some of its nitrogen is absorbed by the rice plant. Part also can be volatilized as ammonia gas or can return to the atmosphere in the form of nitrogen gas as a result of nitrification and subsequent denitrification. In addition, the ammonium ion can be fixed in the clay mineral complex. Nitrogen is absorbed by microorganisms and eventually becomes immobilized in the organic matter. Furthermore, the element can be leached from the soil and so lost in the drainage water. The nitrogen fixed by the clay minerals or immobilized in the organic matter will be partially recovered by subsequent crops.

If nitrogen fertilizer is placed in the root zone rather than applied to the surface of the soil, losses from volatilization and from denitrification can be greatly reduced. The increased efficiency makes it possible to get the same yield of rice from 60 kg/ha placed at a depth of 10 to 12 centimeters as from 100 kg/ha applied in the conventional manner (broadcasting the fertilizer before the last harrowing). Scientists at IRRI have experimentally placed fertilizer inside balls of mud, which then were inserted into the soil between every four hills (a hill is a clump of two or three transplanted rice plants). In 1976 the International Fertilizer Development Center in the United States developed large granules and briquets of urea to substitute for the mud balls and thus eliminate the tedious hand-balling process. IRRI scientists found the briquets to be as satisfactory as the mud balls. The task of placing over 60,000 briquets in each hectare of land by hand is arduous but not at all unfeasible in economies with a surplus of labor. It is no more time-consuming than transplanting rice. Indeed, Chinese farmers in some localities are using the mud ball technique today. It is likely, however, that less laborious methods of fertilizer placement will be devised in the near future.

There is growing interest in finding ways of improving the biological fixation of nitrogen. For example, recent research at

IRRI has shown that when blue-green algae are grown in the presence of the water fern (*Azolla* sp.) the amount of atmospheric nitrogen fixed may increase by a factor of between 5 and 10, as compared with the same concentration of blue-green algae grown alone.

Urea and ammonium sulfate are equally satisfactory nitrogen fertilizers. Because of denitrification losses, nitrogen should not be added in the nitrate form as a basal dressing. It may be used for supplemental dressings after the root systems are well developed and the nitrogen can be immediately absorbed.

Water Management in Lowland Rice

The wide spread of the modern rice varieties since about 1966 stimulated many studies to determine the optimum water management conditions for these short-statured rice plants. Although the appropriate water management practices are influenced by varietal differences and soil conditions, a few general principles can be widely applied.

1. The ideal water depth in the paddy is 5 to 7 centimeters, although depths varying between 2 and 15 centimeters are not harmful. Such water depths suppress weed growth, facilitate the use of granular insecticides and herbicides, reduce losses of fertilizer nitrogen, promote favorable chemical changes in the soil-root zone and, of course, provide a continuous and adequate supply of water to the rice crop.

2. Rice grows best in soil that is continuously submerged from the time of planting until the crop approaches maturity. Any drying and rewetting of the soil not only reduces crop yield but causes losses in soil nitrogen.

3. Under tropical conditions, the growth of the rice crop usually suffers from inadequate water unless rain or irrigation occurs every week or 10 days.

4. When irrigation systems are inadequate or absent and periods of water scarcity occur, the number of days that the drought persists is the factor most highly correlated with yield reductions. However, drought from about 3 weeks before flowering (soon after panicle initiation) to a couple of weeks before harvest is the most damaging. If drought occurs during

the early tillering stage but is followed by adequate rain for the duration of growth, the crop usually is able to recover.

5. When there is an assured water supply, nitrogen fertilizer applications pay off handsomely; and, under most conditions, applications of 60 kg/ha in the wet season and 120 kg/ha in the dry season can be recommended with confidence. In rainfed paddy areas, however, if the likelihood of one or more droughts is high, it is generally unprofitable to apply more than 30 kg/ha of nitrogen.

Chemical Changes in Flooded Soils

Many chemical changes take place when soils are flooded, most of them beneficial to the rice plant. Numerous studies have shown that within the first 3 or 4 weeks after a dry soil is flooded:

- the oxygen supply decreases almost to zero except in a thin layer at the soil surface;
- the pH of acid soils increases, while that of calcareous soils decreases, thus tending to bring most soils to harmless degrees of either acidity or alkalinity;
- iron is reduced from the ferric to the ferrous form, and large amounts of soluble iron are released into the soil solution;
- the supply of available nitrogen, phosphorus, silicon, and molybdenum increases;
- the availability of both zinc and copper decreases; and
- harmful quantities of toxins, such as organic acids, ethylene, and hydrogen sulfide may be produced, and under certain soil conditions the soluble iron quantities may build up to toxic levels.

Of course the last two items are harmful, but the others are distinctly beneficial and advantageous to rice, the only major food crop that thrives in waterlogged soil.

The chemistry of flooded soils is a complex subject, but the most important single change that takes place on flooding is the change in pH. Figure 4 shows the effects of submerging six soils ranging in original pH from as low as 3.4 to as high as 8.2.

pH

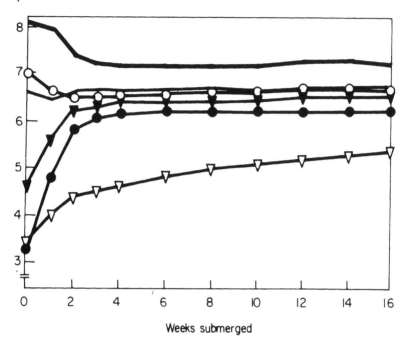

Weeks submerged

Figure 4. Changes in pH of six soils after submergence. (Source: F. N. Ponnamperuma, 1976, *Specific Soil Chemical Characteristics for Rice Production in Asia,* IRRI Research Paper Series No. 2)

After 2 to 3 weeks of flooding, nearly all soils had pH values between 6.0 and 7.2, a suitable range for the rice plant. The one soil that attained a pH value of only slightly over 5 was an acid sulfate clay with a low active iron content and a high acid reserve that prolonged submergence was unable to overcome. The other highly acid soil reached a pH above 6. It had an active iron content well above that of the acid sulfate clay, and thus the quantities of soluble iron released on flooding were sufficient to neutralize much of the soil acidity.

Solar Radiation and Rice Yields

Scientists have shown time and again that there is a high positive correlation between the amount of solar radiation received by the rice plant during the last 45 days before harvest and grain yield. This phenomenon adds greatly to the economic advantage of rice irrigation projects in arid regions and of year-round irrigation projects in those tropical areas

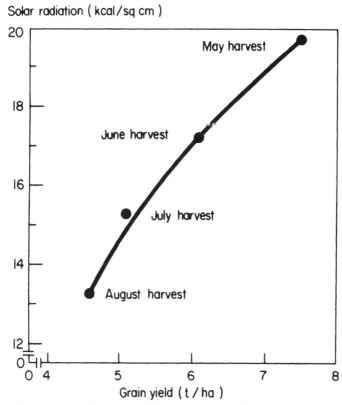

Figure 5.   Grain yield of rice in relation to solar radiation during the 45-day period prior to harvest (average of IR8 and IR5 varieties) in 1968. (Source: *IRRI Annual Report for 1968*)

that have a prolonged dry season.

The relationship between solar radiation and yield is shown in Figure 5. The data were obtained from the date-of-planting studies at IRRI, but the same conclusions have been reached from shading experiments and from the numerous photosynthesis studies carried out in Japan. Over the last 15 years, average dry season yields in IRRI experiments have been about 2 t/ha higher than those in the wet season (see Figure 3).

Although on experimental fields, where insect control is good, the higher yields in the dry season are almost entirely due to greater solar radiation, on farmers' fields the incidence of insect and disease attack is much lower in the dry season than in the wet season, which is an additional factor contributing to the high yields in dry climates or seasons.

Plant Protection

Insect Control

Rice is subject to attack from dozens of insects, and the damage to the plant can be severe. Some of the more common and widely distributed rice insects in Asia are rice stem borers *(Chilo suppressalis, Tryporyza incertulas,* and *Sesamia inferens),* brown planthopper *(Nilaparvata lugens),* green leafhoppers *(Nephotettix nigropictus* and *N. virescens),* white-backed planthopper *(Sogatella furcifera),* the gall midge *(Pachdiplosis oryzae),* and whorl maggot *(Hydrellia philippina).* Other insects that occur sporadically or that are important only in certain locations include the rice bug *(Leptocorisa acuta),* armyworm *(Pseudoletia unipuncta),* rice leaf folder *(Cnaphalocrosis medinalis),* rice hispa *(Hispa amigera),* and rice caseworm *(Nymphula depunctalis).*

Numerous books and publications describe the many insect pests of rice and methods for their control (see Bibliography). No more is attempted here than to indicate the broader aspects of present knowledge of insect control in rice.

Entomologists have concentrated their insect control efforts on developing resistant varieties (working jointly with plant breeders), on the use of insecticides, and on biological control. Regarding the latter method, much has been learned about parasites that attack harmful insects, but there have been no outstanding successes in controlling rice insect populations by rearing and releasing such parasites. Rather, the main achievements so far have been in developing rice varieties that are resistant to insect attack and in identifying the most effective insecticides and devising efficient ways of applying them.

Varietal resistance. The more recent IRRI varieties appear to have medium resistance to the rice stem borers and high resistance to the green leafhoppers and to the brown planthopper (see Figure 2). Unfortunately, most insects have biotypes that vary in their ability to attack a given variety of rice. As the screening and testing of rice varieties and genetic lines became widespread, it was found that certain varieties bred and selected in the Philippines for resistance to a particular insect would prove susceptible when grown in India, for instance. The explanation is that a different biotype of the insect existed in that country. Furthermore, when a new rice variety that is resistant to a given insect is grown over a wide area, in a few years a minor or obscure strain (biotype) of the insect may become a major one, because it is able to attack the resistant rice variety. This propensity of insects to develop distinct biotypes and to shift their population composition complicates rice-breeding programs and makes it necessary to concentrate on discovering sources of broad-based resistance that may be more lasting.

At present varietal resistance to rice stem borers, green leafhoppers, and possibly gall midge as well seems fairly stable. For example, the resistance of such varieties as Peta and IR8 to green leafhoppers has not changed for more than a decade. The brown planthopper, however, has at least three known biotypes. As an indication, IR26, which showed strong resistance to the brown planthopper in the Philippines and elsewhere in Southeast Asia when it was first named, succumbed to brown planthopper attack in the Philippines 2 years later. When tested in parts of India it proved to be similarly susceptible. The reason is that before IR26 was released it had been exposed only to biotype I brown planthoppers, the predominant type in the Philippines and in many other parts of Southeast Asia. Although thus far several IRRI or Philippine varieties (IR36, IR38, IR40, and IR42) show resistance to at least two biotypes of the brown planthopper, there is no assurance that they will continue to do so indefinitely.

The control of the brown planthopper is extremely important, for not only is it the vector of the grassy stunt virus, and of a new virus disease first noted in the Philippines in 1976 called "ragged stunt," but it also causes severe direct damage by feeding on the rice crop. This damage is referred to as "hopperburn."

As indicated earlier, the most promising approach to solving the problem of genetic variations in insect populations is to develop rice varieties with moderate-level, multiple-gene resistance as replacements for the present varieties that owe their resistance primarily to the presence of specific major genes.

Hopperburn in a rice field. Often the insect infestation starts in a small area and rapidly spreads through the entire field of a susceptible rice variety. (Source: IRRI)

In spite of the difficulties involved in breeding rice varieties with resistance to all genetic variants that may develop within insect populations, the use of resistant varieties still remains the single most valuable low-input technology that can raise yields on farmers' fields and, at the same time, greatly reduce the necessity for using costly insecticides.

The use of insecticides. A few years ago many scientists hoped that stable resistance to the major insect pests of rice could be genetically incorporated into a series of high-yielding varieties and thus obviate the need for expensive, environment-polluting insecticides. However, the strong tendency of some insect species to develop biotypes, and the severity of attack by several insect species against which varietal resistance has not yet been found, suggest that chemical insecticides will be needed whenever and wherever insect populations build up to dangerous levels.

Recent studies in several Southeast Asian countries revealed that the chief constraints to high yields on farmers' fields were improper water management, the inadequate use of fertilizer, and poor insect and disease control. The results indicated that the use of insecticides was often uneconomical because of the high cost of the chemicals. Consequently, it behooves the insecticide industry and the entomologists to develop cheaper products and to devise ways of reducing the amounts needed for effective insect control.

In lowland flooded rice, green leafhoppers, brown plant-hoppers, whorl maggots, and stem borers generally can be controlled by systemic insecticides, such as carbofuran, either by spraying the plants or by broadcasting granules into the paddy water. Nevertheless, when populations of brown planthoppers are high, the insect has proved difficult to control. This is especially true when foliar sprays are used, because the brown planthopper feeds mostly at the base of the plant where sprays usually do not penetrate. Furthermore, foliar sprays used against the brown planthopper actually can cause an increase in its population, presumably because they kill parasites and competing insects.

Another complication in the use of insecticides is that insect populations tend to develop resistance against a given insecticide when it is used year after year in the same location. Although the insecticide industry is continually introducing new chemicals, it is barely keeping up with the genetic changes in the insect population.

The most important recent advance in the use of insecticides is the discovery that systemic insecticides are much more effective when deeply placed in the soil. With a simple hand-powered applicator, a liquid systemic insecticide can be placed in the root zone, and one application made soon after transplanting can be much more effective than broadcasting the same insecticide in granular form or than four foliar sprayings (Figure 6). In fact, 0.5 kg/ha of the active ingredient of carbofuran applied in the root zone was fully as effective as 1.5 kg/ha broadcast—the conventional application technique.

Entomologists are testing numerous ways of increasing the efficiency and reducing the cost of insecticide use. They have found, for example, that by dipping the roots of rice seedlings into a mixture of gelatin, water, and a systemic insecticide, the green leafhoppers and the tungro disease (of which they are the vector) can be controlled reasonably well for an entire growing season.

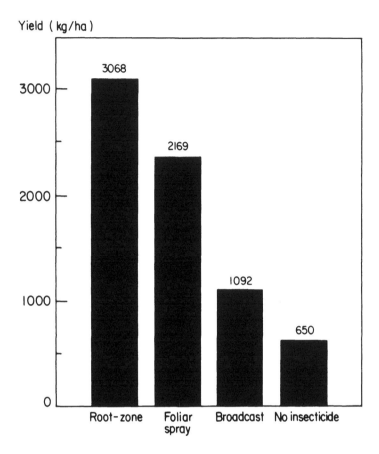

Figure 6. The impact of insecticide placement on yield of transplanted rice. Carbofuran at 1 kg/ha (active ingredient) was applied 3 days after transplanting into the root zone and as broadcast treatment. Monocrotophos was applied four times at 20-day intervals as a foliar spray at 0.75 kg/ha a.i. (Source. *IRRI Annual Report for 1975*)

Integrated pest control—the combination of resistant varieties, management practices, and insecticides—is becoming widely recognized as the most effective and efficient way of keeping insect populations at low levels. For instance, brown planthopper outbreaks are common only where two or more rice crops are grown consecutively in a single year. Thus, planting some other crop between rice crops significantly reduces the brown planthopper population, because the insect has an extremely narrow host range. Transplanting rice seedlings at wide spacing also seems to reduce population build-up. Insect damage to the rice crop undoubtedly can be kept at low levels and yields and profits can be increased by using resistant varieties and by employing such practices as multiple cropping and insecticide placement.

Disease Control

Although there are fungal, bacterial, and virus diseases of rice that can cause severe losses, all of them can fortunately be kept under reasonable control by using resistant varieties and proper cultural practices. Fungicides and other chemicals will cut down the incidence and severity of several important diseases, but their use is uneconomical in the less developed countries.

The most widespread disease of rice is rice blast, which is caused by the fungus *Pyricularia oryzae.* The symptoms are spots or lesions on leaves, nodes, and panicles. They have gray or whitish centers and, although varying in shape and size, are typically elliptical with more or less pointed ends. Rice blast disease forms physiological races, and it is necessary to change rice varieties when a new race breaks out in a locality. So far, plant breeders have been able to create resistant varieties about as fast as new races appear. With the extra attention now being paid to developing horizontal (sometimes referred to as broad-spectrum or multiple-gene) resistance, it seems likely that even greater success in disease control through varietal resistance will occur in the future.

Another widespread fungus disease of rice is brown spot, caused by *Helminthosporium oryzae.* Typical symptoms are oval brown spots scattered over the surface of the leaves and glumes. The disease has been studied thoroughly by plant pathologists, particularly in Japan and India. Nevertheless, if the rice crop is well managed and particularly if it is well supplied with soil nutrients, losses from the disease are not severe.

A third important fungus disease of rice is sheath blight, caused by *Corticum sasakii.* Typical symptoms are greenish-gray, ellipsoid, or ovoid spots on the leaf sheath. In the field the spots usually are observed first near the waterline. They may spread to the upper leaf sheaths and to the leaf blades. When this occurs the entire leaf dies and yield losses are severe. The disease is encouraged by dense planting and by the use of fertilizers. There are only a few resistant varieties, but their use is still the best and most economical method of control.

There are only two important bacterial diseases of rice, bacterial blight (caused by *Xanthomonas oryzae*) and bacterial streak (caused by *X. translucens*). Bacterial blight is the more damaging. Normally, the symptoms of bacterial blight become noticeable in the field as the crop approaches the heading stage. The disease is characterized by lesions on the leaf blade beginning at the edge and enlarging to long yellow areas with a wavy margin. Eventually these lesions may cover the entire leaf blade, and the tissue dies and later becomes infected with saprophytic fungi. In the tropics the disease may assume an additional type (called "kresek" in Indonesia where it was first described) in which young seedlings are attacked after transplanting. The entire plant may die.

Bacterial streak has been found only in the tropics. Typical symptoms are fine translucent streaks that enlarge lengthwise and that later turn brown. In the advanced stages bacterial streak is indistinguishable from bacterial blight.

Scientists have identified varieties that are resistant to the bacterial diseases of rice, and their use constitutes the most practical method of control.

Although there are numerous virus diseases of the rice plant, only four are of economic importance in the tropics. They are the tungro disease, which is transmitted by green leafhoppers; the grassy stunt and the ragged stunt diseases, both of which are transmitted by the brown planthopper; and the hoja blanca disease, transmitted by the planthopper *Sogatodes oryzicola*. Fortunately, none of these diseases can be transmitted mechanically or through the seed.

The symptoms of the tungro disease are stunting of the plants and leaf discoloration ranging from various shades of yellow to orange. Scientists have now found that the "penyakit merah" disease of Malaysia and the "mentek" disease in Indonesia are the same as the tungro disease. The disease occurs sporadically throughout the humid tropics of Asia and can cause severe damage and heavy yield losses.

Plants infected with grassy stunt virus are characterized by severe stunting, excessive tillering, and erect growth. Diseased plants usually produce no panicles. The disease is not yet so widespread as tungro disease and is found mainly in the Philippines.

Ragged stunt is a newly recognized virus disease of rice in tropical Asia. The predominant symptoms are stunting to various degrees at all growth stages and ragged, torn, or serrated leaves. Diseased plants do not deviate markedly from healthy ones in color or in degree of tillering.

The symptoms of the hoja blanca disease are white, chlorotic stripes on the leaves (or even completely white leaves), stunting of the plant, and poor filling of the grains at maturity. This disease occurs almost exclusively in Latin America.

The principal means of control of virus diseases is the use of varieties that are resistant to the insect vector or to the virus, or to both.

Weed Control

During the past decade remarkable progress has been made in developing new herbicides for rice and improving methods of application. It is now possible to control most noxious weeds chemically in irrigated fields, in rainfed paddy, and under upland conditions. Hundreds of experiments all over the world have clearly shown that weed control is essential for high yields. The use of chemical herbicides is not a requirement, but weed control is.

In flooded lowland rice, weed control problems are far less serious than in rainfed paddy or upland rice, because the flood water itself eliminates some weeds and retards the growth of others. Whether to weed by hand or to use chemical herbicides is largely a matter of economics; the cost of herbicides must be measured against the cost of labor.

In the Philippines, for example, many farmers use granular 2,4-D as a pre-emergence herbicide on flooded lowland rice because it is inexpensive and controls most of the annual weeds. At about US$7.00 per hectare, it is cheaper than hand weeding. However, the more selective herbicides, such as butaclor or thiobencarb, cost three or even four times as much as 2,4-D, and under most conditions are more expensive than

Grain yield (t/ha)

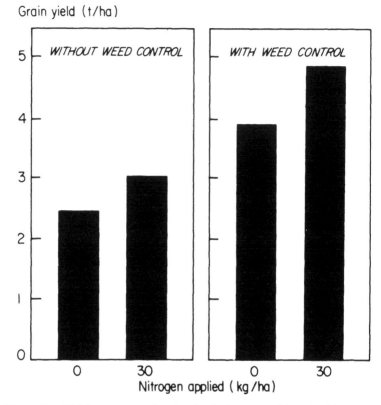

Figure 7. Yield response to low levels of nitrogen, with and without weed control. This is the average of two trials on farmers' fields in the Philippines. (Source: IRRI)

hand weeding. Taiwan offers a contrasting example. In 1975 the average daily wage for weeders there was US$3.75, as compared with only US$0.80 in the Philippines. Thus, in Taiwan the more costly herbicides could be used more profitably than hand weeding.

Weed control does not need to be all chemical or wholly performed by hand labor. Good land preparation, multiple cropping, and straight-row planting (with the subsequent use of the rotary weeder) all aid in weed control. Many farmers use a combination of one application of an inexpensive herbicide (2,4-D) followed by one hand weeding to remove any persistent weeds not eliminated by the herbicide.

An important and basic consideration is that if fertilizers are to be used profitably, weeds must be controlled, for the weeds as well as the rice respond to the fertilizer and compete with the crop. A study conducted in the province of Laguna in the Philippines illustrates this principle well. When weeds were not controlled, 30 kg/ha of nitrogen gave a yield only 0.5 t/ha higher than the yield with no fertilizer. But when weed control was practiced, the yield with fertilizer was 1 t/ha higher than without fertilizer (Figure 7). The actual yields are significant also. On the inherently fertile volcanic ash soil of the Philippines, the yield on unfertilized plots was increased from 2.5 t/ha to 4.0 t/ha by weeding alone; and the top yield obtained from only 30 kg/ha of nitrogen rose from 3 t/ha to 5 t/ha as a result of weed control. There are literally hundreds of experiments showing this same trend. The lesson for the extension worker is that in any program to increase rice yields, weed control comes before fertilizer application when recommending to farmers what practices to improve first.

Although chemical weed control in flooded rice, either transplanted or direct-seeded, is highly successful, it is more difficult to get good control when rice is sown under upland conditions. This is mainly due to the absence of the inhibiting effects of flood water on the growth of weeds and to the fact that some weed species that are difficult to control with chemicals, such as nutsedge (*Cyperus rotundus*), flourish under upland conditions.

Weeding of upland rice fields by hand is expensive and time-consuming. Furthermore, in the early stages of growth (and it is essential to weed rice early), it is difficult to distinguish some grasses from rice plants. Data obtained at IRRI show that some new herbicides suitable for weed control in upland rice fields, even though expensive (US$25 per hectare for chemicals alone), cost less than the two or three hand weedings that are often necessary under upland conditions. Moreover, because of the slowness of hand weeding, the weeding operation often is finished too late to be of value. Use of herbicides, in contrast, eliminates weeds before they can reduce yields.

As herbicides become cheaper (which should be expected as sales volumes increase) and as labor costs rise, chemical control of weeds probably will become a common practice among rice farmers, particularly on farms larger than 2 hectares. Today, however, except in a few regions such as Japan, South Korea, Taiwan, and to some extent the Philippines and the Indian Punjab, few farmers use herbicides for weed control.

Mechanization for the Small Farmer

Although the use of engine-powered equipment on small rice farms is not at all essential for high yields, there are conditions under which mechanization is profitable even for the farmer tilling no more than 2 hectares of land. If labor is either expensive or scarce, it may be more profitable to hire the land preparation and the threshing. Then, too, if a small farmer has the water supply to grow several crops of rice (or of rice and vegetables) in a single year, it may be necessary—and more profitable—to have his land prepared by a power tiller or a four-wheeled tractor, simply because the "turn-around time" is so much less than it would be were he to use animal power. If a farmer tills only 2 hectares of land, he cannot afford to purchase land preparation equipment or threshing machines, unless he either plans to do contract work for other farmers or buys his equipment jointly with neighboring farmers and shares the cost and use with them.

In spite of the nonessential character of mechanization, there has been a steady increase in the number of machines purchased by farmers during recent years. In response to this demand, agricultural engineers have designed machines that are suitable for small farmers and are simple enough to be constructed by the small unsophisticated machine shops scattered throughout the less developed countries.

One of the oldest and most successful of the national programs for developing small machines for rice farmers is in Thailand. This program started in 1953 under the Engineering Division of the newly created Rice Department of the Ministry of Agriculture. Much of the effort was directed toward the design of land preparation equipment, although water pumps and threshing equipment were also developed. Two-wheeled tractors (5- to 9-horsepower) and four-wheeled tractors (12- to 15-horsepower) were designed, tested, and released to local

IRRI's improved power tiller, driven by a 6-horsepower diesel engine drawing a comb harrow. The operator rides on a skid, which saves him from undue fatigue. With a power tiller, preparing 1 hectare of land takes less than 6 man-days compared with 20 man-days with a water buffalo.

machine shops for manufacture. All parts of the machines except the engines were made locally. By 1975 the machine shops of Thailand were turning out 3000 two-wheeled and 800 four-wheeled tractors annually. In addition, the shops made many attachments and implements for larger imported tractors. The locally produced machines and implements usually sold for 30 to 50 percent less than the imported ones.

Taiwan has been successful in improving and locally manufacturing two-wheeled power tillers and other equipment originally designed and produced in Japan.

Since its inception IRRI has had a program to design and test equipment for small-scale rice farming. Three of the machines are described below. The IRRI examples were selected, because information about them is readily available, and not necessarily because they are the only or the best types available.

Land Preparation Equipment

Among the various power tillers imported by the Philip-

TABLE 5. COMPARATIVE LABOR REQUIREMENTS AND COSTS
OF LAND PREPARATION BY POWER TILLERS AND
BY WATER BUFFALOES IN THE PHILIPPINES
IN 1976[a]

	Power tiller		Water buffalo	
	Labor	Cost	Labor	Cost
Item	(man-days/ha)	(US$/ha)	(man-days/ha)	(US$/ha)
Plowing	1.8	24.30	6.6	22.30
Harrowing	3.6	48.60	13.4	45.30
Total	5.4	72.90	20.0	67.60

[a]Based on one plowing followed by three to four passes when harrowing with a power tiller and 9 to 13 passes when harrowing with a water buffalo, which are the ranges encountered in a survey of farmers. Costs based on 1977 value of the Philippine peso: US$1.00 = 7.4 pesos.

pines in the 1960s was one that had a 5- to 7-horsepower motor and was light and sturdy for its weight. In 1971 IRRI engineers designed a tiller based on that machine but simpler, so that, except for the engine and a few bearings and seals, it could be manufactured locally. This model was released to manufacturers in 1972, and by 1976 15,000 units had been manufactured and sold in the Philippines alone.

Surveys conducted in 1975 among farmers and manufacturers familiar with IRRI equipment showed that improvements needed were more power, fewer parts, the addition of steering clutches to facilitate turning, and the capability of accommodating a diesel engine. By 1977, a new prototype that satisfied all those requirements had been produced by IRRI engineers and released to manufacturers in the Philippines and in other countries. This machine, when manufactured in the Philippines, cost about half as much as similar imported machines.

By conducting a survey of 60 farms in the province of Laguna in the Philippines, IRRI engineers and economists were able to estimate the comparative labor requirements and costs of land preparation with power tillers and with water buffaloes (Table

Paddy soil being puddled with a comb harrow drawn by a water buffalo. In Southeast Asia this is the most common method of land preparation for flooded rice. (Source: IRRI)

5). They found that it takes nearly four times as long to prepare flooded rice land with a water buffalo as with a power tiller. Nevertheless, in terms of actual cost per hectare, it is slightly cheaper to use animal power.

Another study showed that in order not to lose money a farmer using a power tiller must cultivate at least 6 hectares of land per year. If the power tiller is to be a good investment, he should use it on a minimum of 10 hectares annually. Therefore, if a farmer owns or has control of 3 hectares of land and if he grows two rice crops a year, he would just break even. The survey indicated that, on the average, power tiller owners actually used their machines on 10 hectares yearly and that, if they did not own enough land themselves, they did contract work on the side.

These studies of land preparation in the Philippines were carried out in areas where lowland flooded rice was grown. The IRRI power tiller has not yet been tested adequately under upland conditions where the power requirements may be considerably higher than on lowland soils.

Direct-Seeding Equipment

Transplanting rice is time-consuming, requiring on the

Direct seeding of lowland rice will become more popular in Southeast Asia in the future, mainly because it can be done with as little as one twentieth the labor needed for transplanting. The IRRI multihopper seeder is inexpensive and is manufactured in several Asian countries. (Source: IRRI)

average 120 man-hours per hectare. It is likely that direct-seeding of rice on lowland rice fields in Asia will become more and more common in the future. Scientists have proved that when water levels and weeds are properly controlled there is no difference in yield between direct-seeded and transplanted rice.

In 1968, IRRI engineers designed and released to local manufacturers a simple hand-operated seeder. It since has been improved, and the current model is called the "IRRI multihopper seeder." The machine is light in weight, has a low center of gravity, and can be built inexpensively in less developed countries. It plants six rows at a time. An operator can sow 50 kilograms of pregerminated rice seed on a hectare of land in about 6 hours. This is 20 times faster than transplanting. A conveniently placed bracket over the wheel allows the operator to lift the machine and turn it when he reaches the end of the field. With its single wheel, the seeder easily can be transported across bunds. The only special requirements for its successful use are good water control, thorough land preparation, and the use of pregerminated seed.

Threshing Equipment

Rice threshing in most of the less developed countries is laborious and inefficient. IRRI engineers have developed and

The IRRI portable thresher is efficient, relatively inexpensive, and easy to move from field to field. (Source: IRRI)

tested a variety of cone, table-type, and drum-type threshers. The most successful model (up to 1977) was the so-called "IRRI axial flow thresher," which will continue to be a good machine for community threshing. It has a capacity of about 1000 kg/hr and can be operated by three or four men. It is rather costly (US$2,000), however, and even three or four average small farmers could not afford to purchase it jointly.

In 1977, IRRI engineers produced a small portable thresher that could be operated and readily carried from place to place by three men. It has a capacity of up to 600 kg/hr of dry paddy and of about 300 kg/hr when the harvested rice is wet. It is powered by a 5-horsepower gasoline engine and weighs 105 kilograms. Its estimated selling price to the farmer is one-fourth that of the axial flow thresher. The design of this IRRI portable thresher was released to manufacturers in 1977. It is anticipated that it will be popular with farmers because of its relatively low price, its durability (it is sturdily built and has few moving parts to wear out), and its portability.

The New Technology and Farm Incomes

After the preceding descriptions of the more important

technological advances made within the past two decades, the question arises as to what benefits accrue to the farmer who adopts the new rice varieties and the improved cultural practices that should accompany them.

Numerous studies have been made to analyze the economic and social implications of the so-called Green Revolution. Among these are a major study by the United Nations Research Institute for Social Development, with information gathered from various localities in the Philippines, Indonesia, Thailand, and Malaysia, and an equally broad study directed by IRRI and published in 1975 under the title, *Changes in Rice Farming in Selected Areas of Asia*. The latter work includes surveys of the impact of the new technology in many villages in most of the countries of South and Southeast Asia. In addition to those two major efforts, numerous investigations have been conducted by individual economists and social scientists in national programs, particularly in India, Bangladesh, and the Philippines.

The results of these studies vary greatly from locality to locality, and to some extent from country to country, because of the diversity of such factors as the physical environment and the infrastructure, including governmental price policies for rice and for inputs. Nevertheless, a few general statements that reflect the impact of the new varieties and management practices can be made.

1. When good water control exists and the modern varieties are properly managed, high yields occur and the new technology pays off handsomely. Obviously, when the new practices are substituted for the traditional, inputs are increased—particularly those of fertilizer, pesticides, and labor (especially as required for better weed control). Nevertheless, when those adjuncts are applied to the modern varieties they are highly profitable, increasing net incomes from 50 to well over 100 percent.

2. With irrigation, the most important single yield-promoting factor is the use of fertilizer—particularly nitrogen, although phosphorus often is required as well. Less frequently

the application of potassium and zinc is also profitable.

3. The second most important input is insecticides. In some localities and seasons, uncontrolled attacks by the gall midge or the brown planthopper, for instance, can devastate a stand and make irrigation and fertilizer use completely ineffective and unprofitable.

4. Despite the heightened use of power tillers and tractors in land preparation and the consequent savings in labor, the total number of laborers hired by those farmers changing from the traditional to the modern varieties has *increased* rather than decreased. In other words, farmers tended to give the modern varieties better care. The extra labor was used largely for transplanting, weeding, and harvesting.

5. Although the situation varies somewhat from country to country, depending upon land reform policies and similar factors, by and large no class of owner or operator—whether landlord, tenant, lease-holder, or owner-operator—tends to benefit unduly from the increased net incomes generated by the adoption of the new technology.

6. Under unfavorable environmental conditions caused, for example, by drought, floods, typhoons, poor weed control, and heavy insect attack, there is usually no economic advantage in growing the modern varieties. In general, the costs of inputs will exceed the increasing income unless a yield advantage of 0.75 t/ha exists.

7. The modern varieties now available are not suitable during the rainy season for many low-lying areas in Asia where levels of water exceed 30 centimeters for extended periods, or where the young rice plants are completely submerged for 10 days or so. Under such conditions of poor water control, farmers tend to revert to growing the traditional varieties. The same farmers, on the other hand, continue to grow the modern short-statured varieties in the dry season when they can control the water level in their paddies.

8. Most farmers adopting the new technology feel that it is beneficial. Except for those with water control problems in the wet season, few farmers have reverted to the traditional

varieties. More studies are needed to reveal the changes in living standards resulting from the increased income provided by the modern technology. In one major study, about 75 percent of the farmers reporting higher incomes indicated that their standard of living also had improved.

Problems of Postharvest Technology

When the rice grain is harvested, it is unusable as human food until the inedible hull is removed. Since the grain is usually consumed as white rice, the bran layer must also be removed (through a process called polishing or milling). Thus, the normal sequence in the handling of a rice crop after it matures is harvesting, cleaning, drying, storage, milling, and distribution to the market (or retention for farm family consumption). Parboiling, if done, occurs sometime before milling. Severe loss can occur when traditional methods of handling are used. Studies conducted in several South and Southeast Asian countries reveal that 13 to 34 percent of the crop is lost during harvest and postharvest operations: during harvesting and threshing, 5 to 15 percent; in cleaning and drying, 2 to 3 percent; in storage, 2 to 6 percent; in processing (parboiling and milling), 3 to 7 percent; and during handling and transport, 1 to 3 percent. Other important losses are grain quality deterioration, under-utilization of by-products, and financial losses due to inefficient postharvest operations. This chapter outlines the nature of such losses and the means by which farmers, millers, and government agencies can increase the efficiency of all phases of rice handling from harvesting to final delivery to the consumer.

This chapter is based on material prepared by James E. Wimberly, rice processing engineer of the International Rice Research Institute.

Harvesting and Threshing

The chief consideration in harvesting is the degree of maturity of the grain, which is determined by measuring moisture content. The optimum moisture content of the rice grain at harvest time is 21 to 24 percent. Under tropical conditions this point is generally reached 28 to 32 days after flowering. If the crop is allowed to stand in the field after it reaches maturity, large losses occur in both the field yield of the crop harvested and the percentage recovery of head (whole grain) rice after milling (Figure 8). If the rice crop is left in the field until the moisture content of the grain is reduced to 15 percent, for instance, the reduction in yield may be as high as 20 percent. The loss is due to a number of factors. A considerable amount of grain simply shatters and falls to the ground before it is harvested; birds and rodents take their share of the ripened grain; and additional losses come about during the harvesting process itself, because the grain is so loosely held on the panicles.

Early harvesting, besides reducing the risk of such losses, produces a higher quality milled rice. When the grain is allowed to remain in the field after it is mature, "sun checking" (cracking of the grain) occurs and many of the grains break during milling. Some farmers object to early harvesting, because the grain is wetter and requires more drying before it can be stored. In addition, threshing early harvested paddy is more difficult. Since many farmers continue to thresh by hand or to drive oxen or tractors over the harvested crop, they are interested in at least medium threshability.

Mechanical threshers remove the drudgery from the process and save time. Although they require a capital investment, the cost of operation is low. Small portable threshers powered by 5-horsepower engines are available and are light enough to be carried readily from field to field (see chapter 2). In some countries large stationary threshers with 25-horsepower engines can be hired.

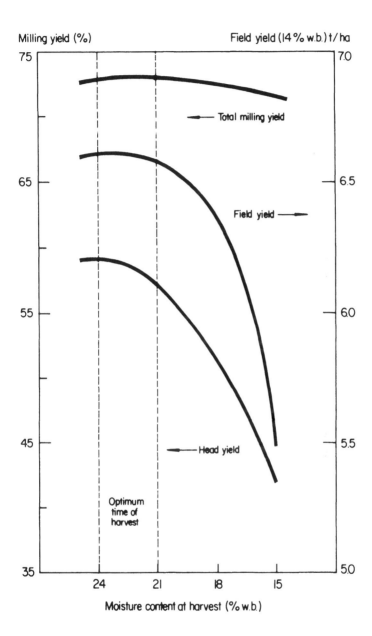

Figure 8 The relationship between the moisture content of paddy (IR8) at harvest and total field yield, the percentage milling yield, and percentage head rice yield. (Adapted from N. G. Bhole et al., 1970, *Paddy harvesting and drying studies*, Rice Process Engineering Centre, Indian Institute of Technology, Kharagpur, India)

Cleaning and Drying

Cleaning

When the paddy is threshed it contains foreign matter, including sand and small stones, straw, and immature and unfilled grains. This extraneous material has to be removed to provide a high quality product. Hand sieving and winnowing are traditional farm methods of cleaning. These methods, however, produce erratic results, and often the paddy sold is of poor quality.

In Asia, where most farms are no larger than 2 hectares, the fact that the farmer cannot afford to purchase equipment for cleaning or drying is a major problem. This is a good reason for having farmer cooperatives that can accept rice from the grower and prepare it for storage and marketing.

Mechanical cleaners of many sizes are available. Fundamentally they employ (1) a vibrating screen (or in some larger mills a rotating screen) with large openings to remove any particles bigger than the rice grain, (2) a second screen with small openings to separate out particles smaller than the rice grain and (3) a blower that forces air through the falling paddy to remove chaff and other lightweight materials. Several less developed countries are building small cleaners suitable for on-farm use. The larger cleaners usually are produced in the developed countries and have to be imported.

The required capacity of a cleaner can readily be determined if the volume of paddy to be handled in a given period is known. Three possible situations are exemplified as follows:

1. Several neighboring small-scale farmers may wish to clean 10 tons of paddy per day. For this they will need a cleaner with a capacity of 2 tons per hour. Such a unit costs about US$400 and can be operated by one man with no extra equipment.

2. A paddy store or mill that receives 1000 tons of paddy every 20 days will need a cleaner with a capacity of about 5 tons per hour. Generally, a bucket elevator is used to lift the paddy to a height above the cleaner. If labor is abundant, however,

A small paddy cleaner suitable for use jointly by several farmers. (Source: IRRI)

workers can carry sacks of paddy up a stairway and dump it into the cleaner from a platform. The cost of a cleaner of the required size is about US$1000.

3. A mill or storage plant that needs to handle 250 tons per day will require a cleaner with a 25 ton-per-hour capacity. A unit of that size is part of a drying and storage complex with mechanical handling throughout the entire process.

Except for farmer-operated units, the selection of cleaners and other processing equipment, and of building facilities, should be part of district and country planning. The 25 ton-per-hour machines would be needed only for an import-export operation in a port or for a large rice-processing complex where at least 10,000 tons are stored annually. The ordinary commercial mill would require a cleaner with a capacity of 5 to 10 tons per hour. The 2- to 5-ton machines would be used primarily by village-level mills and by small cooperatives or groups of farmers.

Drying

The moisture content of paddy is important from the time it is harvested until it is milled. Many farmers and quite a few of the smaller rice mills dry the paddy in the sun on woven mats or on concrete floors. Although this method increases the percentage of broken grains during milling, it is inexpensive and will continue to be a major drying procedure.

Paddy coming from the field usually has a moisture content of between 20 and 24 percent. It should be dried to at least 14 percent as soon as possible to prevent deterioration. Paddy with such moisture content can be stored without much damage for up to 3 months. For longer storage, it should be dried to 12.5 to 13 percent. It is difficult to maintain moisture contents as low as these in the wet season in the humid tropics, where the grain absorbs moisture from the atmosphere.

Some otherwise desirable features of modern rice varieties have made the drying process more complicated. The new varieties often have shorter growth durations than traditional varieties, so they ripen in the rainy season when sun drying is difficult. Furthermore, a number of modern varieties do not have seed dormancy and sprout soon after harvest if allowed to remain wet. To surmount these problems and others, many farmers and millers are purchasing mechanical dryers.

Removing excessive moisture from paddy requires equipment with blowers and supplemental heat. The heat can be supplied from oil, wood, coal, rice straw, rice hulls, or from solar energy collectors. The decision as to what fuel to use depends upon availability and costs in the area where the rice is being dried. In rice mills where the husk is a by-product of milling, it is used for fuel. Most farmers with mechanical dryers use oil, wood, or straw as fuel.

As with paddy cleaning, the capacity of the dryer needed

A bin dryer that holds 1 or 2 tons of paddy is suitable for small groups of farmers or for a village drying and storage operation. (Source: IRRI)

depends on the size of the operation. For farm drying, small 1- to 2-ton bin dryers are used. This type is quite labor intensive, because the paddy is loaded and unloaded by hand. It takes from 4 to 8 hours to dry a batch, depending mainly on the initial moisture content of the paddy. These small dryers, which consist of a wooden, metal, or concrete box with a perforated floor, are simple to construct. A blower is needed to force warm air through the floor and up through the paddy lying on it. Nevertheless, the blower and burner are easy to operate and are relatively trouble free.

Continuous flow systems are used for commercial drying where large volumes of paddy are handled. The paddy enters a tall vertical dryer where the temperature is high, moves on to tempering bins, then back to the dryer, and finally into storage. The flow path of the paddy in such a system is shown in Figure 9. Large dryers such as this are not yet manufactured in most less developed countries. The tempering bins are made of concrete, wood, or metal, their size depending on the capacity needed. The conveying equipment consists of belts and bucket elevators. The initial cost of these systems is high; but if the volume of paddy handled is sufficient, the cost per ton of grain dried is reasonable.

Careful study is needed before investment is made in drying

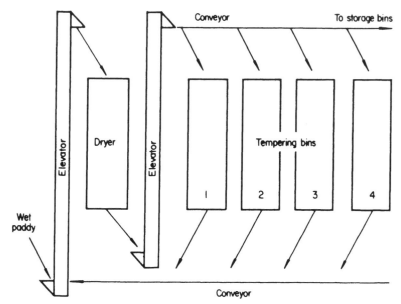

Figure 9. Chart showing the movement of paddy through a modern continuous-flow drying plant. (Source: James Wimberly)

equipment. The size must meet the rice-producing capacity of the district concerned. In some more isolated areas, it is economical for farmers or groups of farmers to use bin-type dryers of low capacity and then to sell dry paddy to the commercial miller or warehouse operator. In other sections, it may be more economical to set up a system in which farmers deliver their wet paddy to governmental or commercial units that have the capacity to dry the paddy well and quickly.

Gains from Proper Cleaning and Drying

For maximum profits and minimum losses it is important for both farmer and buyer to handle good quality rice, free of foreign matter and dried to a moisture content no higher than 14 percent. But most small farmers do not have mechanical cleaners and dryers, principally because their rice output is too small to justify expensive equipment.

There appear to be only two solutions to the problem. One is for the farmers to form small groups or cooperatives, thus gaining enough capital and volume to be able to purchase the cleaning and drying equipment. The other is for either commercial buyers or government-owned warehouses to accept the many small lots of grain from the surrounding area and to clean and dry them.

Understandably, the commercial buyers or the government purchasing agencies have to adjust the price paid to the farmer in relation to the amount of moisture and foreign matter in the paddy he sells them. It is important that these measurements be made accurately and that a fair price be paid for the rice. Reliable moisture meters and dockage testers are available and should be employed. In the more sophisticated markets, the buyer may wish to measure grain size and shape, to keep varieties separate (when a premium is paid for certain varieties), and even to determine the potential milling yield by putting small samples through laboratory shellers and polishers.

Good relations and mutual trust between the farmer and the rice buyer (whether private or governmental) are essential. They can be maintained only if paddy grades and standards are followed. It is up to the governmental agencies that enforce the regulations to see that both seller and buyer are protected.

Purchasing and pricing systems vary from country to country; no single description can cover them all. However, the economic gains that can result from producing clean, dry paddy can be shown by a hypothetical example:

The government has a guaranteed minimum purchase price for paddy of $125 per ton provided it contains no more than 14 percent moisture or 1 percent foreign matter. Government policy reduces the price paid to the farmer by $1.00 per ton for each 1 percent of moisture above 14 percent and by $0.50 per ton for each percent of increase in foreign matter above the 1 percent allowed. Assuming 200,000 tons of paddy are marketed annually and that the average moisture content of the rice is 16 percent, rather than 14 percent, and the dockage (foreign material) 3 percent instead of 1 percent, then the amount of money paid to the farmers would be $600,000 less than if the paddy had met the established standards. Rice processing specialists estimate that moisture meters and rice cleaners to handle that volume of rice could be purchased for $50,000.

While there are other factors to be considered than those pointed out here, nevertheless it generally is to the farmer's advantage to do the best job he can in drying and cleaning his paddy before he delivers it to the buyer.

Handling and Transportation

Studies show that from 1 to 3 percent of the paddy can be lost during handling and transportation. The more times the paddy is handled and moved, the greater the losses. Furthermore, the cost of handling and transporting paddy can vary by several hundred percent depending on the system employed.

To use an extreme example, in some countries bagged rice may be handled as many as twenty-three times and moved by truck three times from when it is delivered by the farmer to a purchase center to when it finally reaches the consumer. From the purchase center, it goes to a paddy store and into storage. Three months later it is transferred to a small mill that has no belt conveyor and is hand carried from cleaner to sheller to separator to polisher. Then it is trucked back to the rice store and finally to the retail market. In the course of its handling and traveling, it is weighed three times.

To consider the other extreme, the farmer delivers unbagged (loose) grain to a modern rice handling and processing plant. It is dumped into a receiving bin and moved by conveyors through the cleaning and drying processes and into storage. Some three months later it moves on a conveyor belt into the rice mill where it is dehulled, polished, and graded in a continuous operation. Finally it is automatically bagged, loaded onto a truck, taken to the retail market, and moved to the sales counter. Instead of being subjected to twenty-three operations, the paddy in this second instance is handled only six times by mechanical conveyors, twice by hand, and once by truck.

The traditional system is labor intensive, whereas the modern one is highly mechanized. Studies show that the modern rice handling and processing plant, provided it runs to capacity, is more economical than are traditional methods in terms of cost per unit of rice outturn. Moreover, the physical losses of grain are reduced substantially.

Here are some practical suggestions for reducing the losses and costs during the handling and transportation of paddy. (Some points, so obvious as to be understood without mention, are included solely for completeness.)

1. A detailed study can be made of the present system to see how the number of handling operations might be reduced.

2. An inventory of existing transport facilities can be made with the aim of using all equipment to the maximum degree possible.

3. Because transporting bulk rice is more efficient than transporting bagged rice, existing truck bodies can be converted to accommodate bulk grain.

4. Torn jute or plastic bags should be repaired or replaced.

5. Farmer-operated tractor-trailers can be used for hauling grain from farm to processing plant.

6. Storage and processing plants should be adjacent to avoid unnecessary transportation time and cost.

7. Even small village mills may find it profitable to install some mechanical devices (such as small bucket elevators to hoist paddy to the grain cleaners).

Storage

Large losses of paddy and milled rice occur during storage. Fortunately, these losses can be virtually eliminated. The construction of storage facilities represents the largest single investment in the postharvest industry.

Physical losses in storage range from 2 to 6 percent. They are chiefly the result of insect and rodent damage and of careless handling, including improper care of the sacks. Furthermore, fungi and bacteria may reduce quality if the relative humidity of the air in the storage space remains too high (above 70 percent), or if rain water is allowed to enter the storage building.

The cardinal rules for good storage are to allow only clean and properly dried paddy to enter storage and to keep the storage building completely rainproof. In addition, the floor should be tight so that no moisture seeps upward through it. In bag storage, wooden dunnage (platforms) should be used to maintain an air space between grain and floor.

Types of Storage Facilities

The principal factor determining the size of the storage building is plainly the amount of paddy to be placed on the market from the nearby area. The building material selected can be reinforced concrete, brick, wood, or sheet metal, according to local availability and cost. Even the size of the building will depend upon local construction skills. In some places it may be better, for example, to construct two 2500-ton storage buildings rather than one structure holding 5000 tons.

Storage facilities are of two types: those accommodating grain in bags and those built to store loose paddy (referred to as bulk storage). Bag storage buildings are usually square or rectangular while bulk plants can be either rectangular or in the form of tall, round silos. Bag storage is labor oriented; bulk storage plants require a larger investment but are less expensive to operate. Losses during storage are not influenced particularly by whether the paddy is stored in bulk or in sacks. Rather, they are determined by the care taken to prevent loss and deterioration.

Cleaning and drying operations may be attached to either type of facility. However, from the standpoint of economy it is especially important to have cleaning, drying, and storage in bulk storage plants, because a system of mechanical conveyors can move the grain from one place to another with minimum handling.

To provide an example of the relative costs of bag and of bulk storage, the costs for 6000-ton storage facilities of the two types are shown in Table 6 (these are figures from one country and reflect only a particular local situation at the time they were assembled). The data simply support the fact that a greater initial investment is required for bulk storage plants and that those using jute or plastic bags are more expensive to operate. Nevertheless, bag storage facilities may be the wiser choice in labor-surplus economies.

Providing Additional Storage Capacity

As rice production goes up, storage space should increase

TABLE 6. A COMPARISON OF THE INVESTMENT AND
OPERATIONAL COSTS OF 6000-TON BAG AND
BULK STORAGE FACILITIES.

Item	Bag storage	Bulk storage
Initial investment		
Building construction	US$150,000	US$150,000
Mechanical equipment	–	60,000
Annual operating costs		
Staff salaries	520	520
Labor	7,200	600
Plant maintenance	1,500	4,500
Power	–	500
Depreciation	7,500	13,500
Interest on investment	15,000	21,000
Purchase of bags	20,000	–
Total operating cost	$ 51,720	$ 40,620
Operating cost per ton of paddy	$8.62	$6.77

acccordingly. Furthermore, most rice-producing countries do not have the facilities to store the paddy in years when harvests are exceptionally good or to accumulate buffer stocks for any lean harvest years in future.

The additional storage facilities should be built in the areas where the rice is being produced and near the processing plants, thereby saving transportation costs. Paddy rice should not be moved long distances. Milled rice can be transported for one-half to two-thirds of the cost of paddy.

Future storage needs can be estimated through a detailed analysis of present capacity in relation to the amount of paddy available for storage. Then, by estimating the increased production that is likely during the next decade or so, an annual storage building program can be prepared. The needed storage expansion will vary widely from country to country— from as little as 5 percent to as much as 75 percent for the decade ahead.

Few countries are prepared for emergency storage in times of exceptionally high harvests. Temporary storage can be provided outdoors by stacking the paddy in bags on wooden dunnage and covering it with plastic sheets or waterproofed canvas tarpaulins.

Losses in Storage

The major causes of storage losses of paddy are birds, rodents, insects, and attack by microorganisms when the moisture content of the grain is too high. Fortunately, these losses can be reduced to insignificant levels by following known storage management practices.

Bird-proofing is accomplished by placing wire screening over openings under the eaves and gables and over ventilators or windows that remain open. Making a building impenetrable by rats and mice is more difficult and costly. Broken floors, doors, and low windows are the most frequent avenues of entrance by rodents. Floors can be repaired or replaced, rodent-proof doors and window frames can be installed, and metal shields can be placed around the base of the storage building. In addition, rat poison can be used within the building.

Fumigation is the only practical way to control insect damage. Jute sacks often contain insects and should be treated with chemicals or fumigated. When an entire building is to be fumigated it must be temporarily sealed to prevent the escape of the fumigating gas. The technique of fumigation is well known, however, and operators can be easily trained to perform the task.

The control of the moisture content of the air, to prevent damage by bacteria and fungi, is more difficult than is controlling insects. In the humid tropics the moisture content of the air is often high enough to cause previously dried grain to increase in moisture content. For example, if the relative humidity of the incoming air in a storage warehouse is 90 percent and the air temperature is 27°C, the equilibrium moisture content of the grain will be 17 percent—some 3 to 4 percent higher than a safe storage level. On the other hand, if the paddy placed in storage bins has a moisture content of less than 14 percent and if the incoming air has a relative humidity of not over 70 percent, there will be no problems with fungi or other microorganisms attacking the grain. Nevertheless, in bulk storage it is essential that the storage compartments are continuously aerated by forcing a stream of air through the grain.

Rice Processing

Parboiling

The parboiling of rice has been practiced in some countries, notably India and Sri Lanka, for hundreds of years. In recent years it has gained some popularity in the United States and a few European countries. Perhaps 12 percent of the world's rice crop is parboiled.

Parboiling consists of soaking the paddy, then steaming it, and finally redrying it before milling. During soaking and steaming the starch swells and becomes gelatinized. When the grain dries, the endosperm hardens and becomes resistant to breakage during milling. Parboiled rice has a different taste and texture than rice that has not been parboiled. Those accustomed to it seem to prefer it. Those who are used to eating the unparboiled product do not take readily to parboiled rice.

Parboiling rice has several advantages. First, dehulling is easier, because the husk is split during parboiling. Second, the extra strength acquired by the kernel during parboiling reduces the number of broken grains during milling. Third, because parboiled rice is harder it tends to resist insect attack during storage. Fourth, the loss of solids into the gruel during cooking is less in parboiled than in raw rice. Fifth, parboiled rice withstands overcooking without becoming pasty. Sixth, the bran from parboiled rice contains from 25 to 30 percent oil, while bran from raw rice contains only 15 to 20 percent. Moreover, the oil from parboiled rice bran, because it has a lower concentration of free fatty acids, is of superior quality.

There are also several disadvantages. First, the heat during parboiling destroys antioxidants, so parboiled rice becomes rancid more easily than raw rice during storage. Second, parboiled rice takes longer than raw rice to cook to a given degree of softness. Third, parboiled paddy must be redried before milling—an additional cost. Fourth, parboiled rice is

Parboiling tanks in a modern plant. (Source: James Wimberly)

harder to polish than raw rice; hence, milling is more costly in time and power. Fifth, the parboiling process requires a large extra investment in capital equipment and operation costs.

Traditionally, parboiling consists of soaking the paddy in unheated water in concrete tanks and then steaming it in metal tanks. It is then dried in the sun on a concrete floor. The process takes from 24 hours to several days, depending on the degree of sunshine. In less developed countries, to reduce the costs of parboiling, rice husks often are used as fuel to fire the boilers producing the hot water and steam and to dry the rice after parboiling. Because of unsanitary conditions during soaking, rice parboiled by the traditional method is a different and inferior product to that treated in a modern parboiling plant in, say, the United States or Europe.

The more modern method is to soak the paddy in hot water and then to steam it. The rice is kept free of foreign matter, and the use of hot water for soaking prevents deterioration. In less developed countries, the paddy is usually sun dried after parboiling, though in a few locations mechanical driers are used. In developed countries, all of the paddy after parboiling is dried by putting it through a continuous-flow, mechanical dryer. Although modern equipment costs more than the traditional, its larger capacity actually reduces the cost per ton of processed paddy.

Rice Milling

Rice milling is the process of removing the outer husk and all (or part) of the bran layer from the grain. The husk is totally inedible. If the bran layer is not removed, the product is called brown rice. Although brown rice is available on some markets, it is not popular because it tends to cause digestive disturbances. Furthermore, the oil in the bran layer is likely to get rancid, especially in hot climates. In the more developed countries particularly, almost all of the bran layer is removed to give a highly polished white rice that is preferred by the market.

There are three principal types of rice mills now in use. One is the steel huller, a rather simple machine that removes the husk and bran in one operation. Its defects are that it is expensive to operate, has a low capacity and a low rice outturn, and produces too many broken grains. A second type is the sheller mill. It consists of several machines: usually a cleaner, a disc sheller, a separator, and a polisher. It has a high capacity and medium operational costs but is less efficient than a modern mill. The modern rice mill has highly efficient equipment: cleaners remove foreign matter; rubber roll shellers dehusk the grain; paddy separators remove any unhulled grain; polishers remove the bran layer; and graders separate broken grains from head rice. The modern rice mill has a high capacity, and it recovers more total rice as well as head rice. A diagram of the sequence of processes in a modern rice mill is shown in Figure 10.

The recovery of polished rice in a mill is termed "percentage outturn." The outturn is affected by variety, by whether the rice is raw or parboiled, by rice moisture content, and by the type of mill used. If rice of the same variety and condition were put through the three types of mills described, the average outturn would be 70 percent for the modern mill, 68 percent for the sheller mill, and 64 percent for the steel huller mill. A further

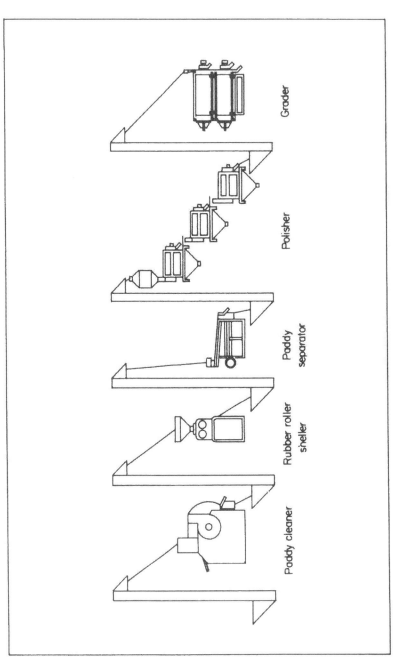

Figure 10. Basic design of a modern rice mill. (Source: James Wimberly)

TABLE 7. AVERAGE RECOVERY EFFICIENCIES OF THREE
TYPES OF RICE MILLS

Type of mill	Recovery as percent of paddy		
	Bran plus hulls	Whole grains	Broken grains
Steel huller	36.6	46.5	16.9
Sheller mill	32.5	55.9	11.6
Modern mill	30.0	62.0	8.0

breakdown into bran plus hulls, head rice, and brokens is given
in Table 7.

To be economical, modern rice mills require trained
operators and a higher level of management than is needed for
an outdated mill such as the steel huller. However, local
personnel can be trained to run the modern mills. Some studies
have been run on the economic benefits of the modern mill, the
sheller mill, and the steel huller. Based on a 1-ton-per-hour
capacity and on milling a total of 6000 tons of paddy annually,
the modern mill gave an increased profit of about US$1.75 per
ton of paddy over that of the sheller mill and of nearly US$5.00
per ton over that of the steel huller.

It should not be inferred that every country should replace all
of its traditional steel huller mills with modern types.
Although the modern mills are more efficient to operate if run
at full capacity, there are other considerations. In Indonesia, for
example, a study by C. Peter Timmer shows that with a scarcity
of capital funds and a lack of alternative employment
opportunities for unskilled labor, the labor-intensive steel
huller mills have substantial economic and social advantages.
There are numerous examples in South and Southeast Asia of
large modern rice mills now being run at a loss because of an in-
ability to provide enough paddy to operate them at full capacity or because of poor management. In many countries with
scarce foreign exchange, it is better to upgrade sheller rice mills
than to purchase modern mills.

Rice Mill Capacity

Modern rice mills are available in capacities ranging from 1
ton per hour to over 10 tons. The larger the mill, the greater the

efficiency, provided enough paddy is available to run the mill full time. A 1-ton-per-hour mill requires nearly the same manpower as a mill with a capacity of 4 tons.

Modern rice mills are designed to operate almost continuously for up to 300 days a year. In determining how many hours a day to operate a mill, factors to be considered are the available labor and supervisory personnel for a three-shift-per-day operation and the availability of paddy, including transportation facilities. Most mills are more economical to operate with one shift per day for 300 days than two shifts scheduled daily for 150 days.

In most less developed countries, because both government and private milling operations exist, there is an abundance of milling capacity. The bottlenecks in the postharvest industry are more likely to be the drying and storage facilities.

Rice By-products

Milled rice, the final product of rice processing, represents only 65 to 70 percent of the original paddy. The products removed, hull and bran, have several uses.

Hulls. The husk or hull of the rice grain, which represents on the average 22 percent of the weight of the paddy, is high in silica and lignin, but it has low feeding value for animals, and none for human beings. Its principal use is as fuel for parboiling and drying.

Hull-fired boilers are high in price, but the fuel is so inexpensive (having few other uses) that it costs little to run the heating plant once it is purchased. For example, the energy value of rice hulls is lower than that of either coal or diesel oil; but in terms of kilocalories per dollar invested in fuel, rice hulls are by far the least expensive (Table 8).

Hulls are also an economical fuel for direct-fired furnaces to produce hot air to dry paddy. Or rice mills that do not parboil can use rice hulls to produce steam to power the mill. Yet, because of the high cost of boilers and steam engines, it is often more economical to run the machinery with electric motors, provided that electric power is available. Some bran oil extraction plants use the hulls as fuel to produce the steam necessary for the extracting process.

Rice hulls also have some specialized uses. The white ash

TABLE 8. ENERGY VALUE AND COST OF FOUR SOURCES
OF ENERGY IN INDIA

Fuel	Energy value	Cost (kcal per US$)
Fuel oil	10,240 kcal/liter	60,000
Coal	5,560 kcal/kg	350,000
Electricity	860 kcal/kwh	43,000
Paddy hulls	3,300 kcal/kg	2,300,000

produced from burning them is sold as a cleaning compound for floors and for absorbing grease. It also has limited use as fillers, additives, and carriers in the chemical industry. Rice hulls can be used as an aggregate for concrete blocks, as a base for pressed board products, as cattle feed and litter, as an additive in potting soil used by nurserymen, for the manufacture of furfural (a chemical product used in the dye and plastics industries), and for insulation. In some of the more developed countries, hulls are compressed into briquets and sold as a firewood substitute.

Rice bran. There are two principal uses for rice bran: as a feedstuff for cattle, poultry, and swine, and as a source of rice bran oil. About 8 percent of the weight of the paddy is bran. Compared with the rice endosperm, it is rich in vitamins, minerals, and protein. The demand for bran as animal feed is sufficiently high that many small rice mills will process grain for farm families free of charge if the mill is allowed to retain the bran, which it sells to manufacturers of feed.

Commercial bran polish, which includes the germ, contains from 15 to 20 percent oil. But the bran of parboiled rice may contain as much as 35 percent oil. If properly processed, rice bran oil is comparable to other vegetable oils for cooking, for salads, and for shortening. Lower quality rice bran oil is used in soap making and for a few other industrial purposes.

Other By-products

Broken rice, which is not readily salable in the more sophisticated markets, is used for manufacturing beer and wine. Some West African countries, however, import broken rice grain to mix with local rice, thus providing consumers with a low-cost staple.

In developed countries, processed rice products such as breakfast cereals are in demand. Rice flour can be used as a partial substitute for wheat flour in bread making. However, because of its low gluten content, it cannot constitute more than 30 percent of the flour mixture. Rice flour is also used in baked goods for those who are allergic to wheat flour and to other cereal grains. Little rice flour is produced, however.

Rice Distribution and Quality Control

Distribution

Because milled rice deteriorates much more rapidly than paddy, milling takes place only a short time before the rice is placed on the market. Even then, special care must be taken to preserve the quality of the product in the interval between milling and consumption.

Milled rice is generally stored from 2 to 3 weeks to as long as 2 to 3 months. As with paddy, milled rice must be protected from moisture, birds, rodents, and insects. Although it can be kept in loose bulk storage, milled rice normally is placed in the jute or plastic bags in which it will be marketed. The bags should be new or properly reconditioned before use. Wooden dunnage always should be used to keep the sacks of rice off the floor of the warehouse.

Because the demand for rice in the heavy consuming countries is rather constant throughout the year, the most economical way to handle the product is to schedule pickup and delivery so that rice will flow steadily into and out of the warehouses. If a steady flow is established, fewer transport vehicles will be needed than in a haphazard distribution system.

To plan an efficient rice distribution system it is necessary to determine (1) the amount of rice needed at all major consumption points, (2) the amount of rice available in the area being analyzed, and (3) the transportation facilities needed to move the rice. In most rice-growing countries, wholesale rice dealers and their storage facilities, as well as those of the government, are located in the larger district centers. Thus much of the rice must be moved from the smaller villages to the larger centers and then again to the urban retail markets. In some areas, though, systems have been developed to take the rice from the locality in which it was grown directly to the retail market, thus saving extra transport and handling costs.

Quality Control

Quality standards in the rice markets of South and Southeast Asia (with the possible exception of Thailand) are not so high as in the more sophisticated markets of the United States and Europe, for example. Consumers in Asia tend to select their rice by appearance, generally preferring slender grains free of chalkiness. In some areas, the buying public knows rice varieties and asks for them by name, in order to be sure of getting rice with a preferred eating and cooking quality. The quantity of broken grain is also an important factor. Many people are willing to pay a premium for rice that has no more than 10 percent broken grain. Rice customers everywhere naturally prefer clean rice that is free of foreign matter and damage by insects or fungi. In an effort to standardize national rice grading systems, the FAO's Intergovernmental Group on Rice developed a model grading system for rice in international trade. The latest revision was distributed to all interested governments in 1972.

It appears that the only countries that have adhered to rather strict grades are those such as Thailand that have a sizable export market that demands certain quality standards. In Thailand, a private organization, the Rice Traders Association, has established and now polices the system of standards. The government supports the association's program, but it did not take the initiative to set up the system originally.

With increasing affluence, the local markets of Asia probably will become more quality conscious than they are now. It is of prime importance that minimum quality standards be enforced to protect both buyers and sellers.

The Systems Approach

Too frequently countries use a piecemeal approach to improving their postharvest operations. That is, in one

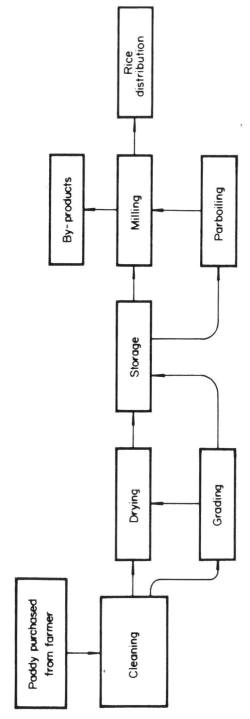

Figure 11. Sequence of postharvest operations. (Source: James Wimberly)

program the nation attempts to better its storage facilities, in another program its drying processes, and so on. Although such efforts are beneficial, greater savings are possible when the entire postharvest system is attacked.

A simplified flow diagram of postharvest operations is shown in Figure 11. A systems approach to cutting losses and improving efficiency includes (1) matching the cleaning and drying facilities to the purchasing program, (2) adjusting the storage capacities to the receiving and milling schedules, and (3) matching the milling capacities and facilities to the storage system and to the rice distribution requirements. That sort of approach provides maximum utilization of existing facilities, minimum investment in new facilities, and the lowest possible operational cost. A study in Sri Lanka showed that when the scheduled systems approach was applied to the entire postharvest operation (as compared with the piecemeal approach), the nation's processing costs fell by US$7.30 per ton of paddy.

The size of storage and rice processing plants is a major consideration. The economics of three plants of varying capacity is shown in Table 9. It is evident that larger mills cost more but that the profit per ton of paddy is greater. Moreover, the number of employees per ton of paddy is lower in the larger mills. These figures, obtained in Sri Lanka, reflect the situation when all three mills were run at full capacity. Whenever they were operated below their rated level, the cost per ton increased.

In summary, to take advantage of the economics of the systems approach the following steps can be helpful:

1. analyzing the present status of the rice industry in a district or region, including making a study of procurement programs, of cleaning and drying facilities, of storage and processing capacities, and of problems in transportation and distribution;

2. becoming familiar with the newest technology and available equipment;

3. taking into consideration each step shown in Figure 11 when planning the improved system;

4. attempting to remove barriers to implementation and making policy changes when necessary;

TABLE 9. THE ECONOMICS OF THREE SIZES OF MILLING
OPERATIONS. (Data from a recent study in Sri Lanka;
values in US $)

	Processing capacity of mill (tons of paddy per year)		
	6,000	12,000	24,000
Investment cost (thousands)	$465	$720	$1100
Annual operating cost (thousands)	83	131	192
Operating cost per ton of paddy	14	11	8
Profit per ton of paddy	8	11	13.50
Number of employees	31	53	71
Number of employees per 1000 tons of paddy	5.16	4.42	2.96

5. encouraging and supporting the local manufacture of modern postharvest equipment;

6. setting up training programs for both management personnel and semi-skilled employees; and

7. reviewing the system from time to time and making any necessary changes in line with current technological developments.

In conducting surveys using the systems approach, a country may find it necessary to obtain expert assistance from outside agencies (see appendix). Such studies require the participation of engineers, economists, and marketing and management experts. This systems approach is obviously a government task. Nevertheless, a number of specialists in the economics of rice handling and marketing believe that governments should do no more than help remove the bottlenecks and should let private industry make the major investments in rice processing and marketing. The decision will depend considerably on the kind of political and economic system that exists in a given country. For instance, Thailand and Sri Lanka might act quite differently in improving their postharvest systems.

4
Rice Marketing

In Asia, and to an extent in Latin America and Africa, the marketing of rice has several conspicuous features. First, supply and price fluctuate sharply. Although demand is steadily increasing in line with population growth, supply—and consequently price—may vary greatly from year to year, depending upon weather conditions. Second, on-farm rice consumption is heavy. From 50 to 70 percent of the rice crop in the less developed countries of South and Southeast Asia is consumed by farm families and never reaches either local or international markets. Third, exports are limited. Only about 4 percent of the world's rice crop enters into international trade. In other words, on the average 96 percent is consumed in the country in which it is grown. Fourth, crop surpluses are improbable. The countries of Asia, as a whole, approach self-sufficiency in rice only in years of favorable weather. Therefore, there is little likelihood of any sizable accumulation of excess rice in Asia as populations continue to increase. Finally, modern rice marketing techniques are needed. In most less developed countries, rice marketing policies, drying and storage facilities, and systems for the control of market supplies and prices require substantial improvement. Of the foregoing characteristics of rice marketing, only rice marketing techniques are likely to change significantly (at least in Asia) during the rest of this century.

This chapter is based on material prepared by J. Norman Efferson, chancellor, Center for Agricultural Sciences and Rural Development, Louisiana State University.

The subject of rice distribution and marketing, including pricing policies, is a complex one and deserves more thorough treatment than can be given here. Nevertheless, a few major factors that administrators should keep in mind can be listed.

Local Marketing

A sizable segment (as much as 50 percent in some countries) of the small farmers of South and Southeast Asia are still at the mercy of the rice buyer, in spite of a trend to improve the marketing of rice through such measures as stabilizing prices, regulating the practices of rice dealers, and improving collection, drying, milling, and storage facilities. Too frequently, the farmer has to accept the buyer's offer, because he needs immediate income to support his family. In addition, he is likely to be in debt, perhaps to the same person (the moneylender) who offers to buy his rice. Furthermore, most small farmers in Asia do not have satisfactory storage facilities for holding any crop surplus until prices rise after the harvest season. Another factor contributing to the plight of the small rice farmer is his isolation from the marketplace because of inadequate roads. Inaccessibility discourages farmers from producing more rice than they need for subsistence.

Although some countries have established policies that not only guarantee a satisfactory minimum price to the farmer but also stabilize prices from year to year, there is great need for further improvement. Most countries do not have enough storage capacity to accumulate buffer stocks in times of bumper harvests. Adequate storage is a requirement both for price stabilization and for protection against importation in lean harvest years.

The lack of well-run cooperatives or farmers' associations in many rice-deficient Asian countries also contributes to the adversity of the small farmer. Efforts to establish cooperatives among farmers in Southeast Asian countries have had limited success. Often the cooperatives are run by poorly trained managers, the objectives and the advantages of the cooperatives are not well explained to farmers, and farmer members are given little opportunity to participate in the decision making.

If those defects are rectified, and if the cooperatives are run by honest people dedicated to the farmer's well-being, such organizations can aid immensely in marketing and in supplying inputs at reasonable prices.

Nevertheless, some improvements in the handling and marketing of rice are being made. Most countries are taking steps to establish pricing policies, to build new drying and storage facilities, to modernize rice mills, to organize farmer cooperatives, and to improve credit systems. But the job, which is being done slowly, needs to be intensified and accelerated. In most nations, the large capital investments involved in these improvements requires that foreign aid be sought to finance the projects.

In many countries, government agencies have been established to handle rice pricing, purchasing, and selling. Examples are the National Grain Authority in the Philippines, the National Logistic Authority in Indonesia, the Padi and Rice Marketing Board of Malaysia, the Paddy Marketing Board of Sri Lanka, the Union of Burma Agricultural Marketing Board, and the Food Corporation of India. Fundamentally, each such organization sets policies aimed at providing adequate supplies of rice in the nation, regulates imports and exports, and enforces whatever pricing policies have been decreed by the government. To the extent that such agencies are free of corruption and unfavorable political influence, they are an advantage to any country. Where rice is the staple food of a nation, they are in fact a necessity, since some government agency has to have the authority to control the rice industry.

With the exception of Thailand, Burma, and Pakistan, and possibly Vietnam, Laos, and Kampuchea, it is doubtful that any country in South and Southeast Asia will be a steady exporter of rice in the foreseeable future. Consequently, the primary concern of most governments in those regions is to develop pricing policies and marketing facilities that encourage maximum production at the most economical levels. Undoubtedly, a number of Latin American and African countries will become important exporters of rice as they develop their rice industries and as the world demand for rice increases.

Self-sufficiency as a Goal

During the serious transportation problems and input shortages of World War II and the early post-war period, many rice-consuming nations were critically short of their principal source of nourishment. Prices soared, and consumers were forced to eat many less desirable foods in order to survive. Elections were won and lost depending on the volume of rice in local markets and the prices at which rice could be purchased. Resolutions were made and laws passed to try to ensure that the situation would never occur again. In some countries, buffer-stock programs and price support policies were introduced and have been continued since that time. The goal of practically all of the deficit rice-consuming nations of the world has been self-sufficiency.

In the three decades since then, the global demand for rice has continued to increase at 2.5 to 3.0 percent annually. World production of the grain has risen at about that same rate, but output from year to year has been erratic. In some years, total production increased twice as fast as demand. Consequently, rice being plentiful, prices in the next year were relatively low—much lower than the internal support prices in some countries. At other times, the production level was less than that of the previous year, while demand continued to increase. After such a year, rice prices went up drastically around the world, and in some deficit countries adequate rice supplies at reasonable prices were almost impossible to obtain.

The outlook is for a steady increase in demand for rice and a continued expansion in production equal to demand but with widely variable production from year to year, mainly due to weather. As in the past, in some years there will be surpluses and low prices, and in others, shortages and high prices.

The rice-growing countries have to decide whether to aim for self-sufficiency and at what cost. The answer is tied to conditions in the country under consideration. If a nation—such as India—has a foreign exchange problem, a current low average yield of rice and a high domestic requirement, it certainly is justified in doing everything possible to boost production. Nevertheless, "self-sufficiency at any cost" is not a valid goal for every government. Certainly, rice prices should be stabilized at levels high enough to encourage farmers to seek better yields. Necessary, too, is a favorable ratio between the price of rice and the cost of inputs the farmer needs for obtaining good yields. In some countries the price at times was set so high that excess production was stimulated and the government had to dispose of the surplus at a loss. This has occurred in Japan where rice prices have been double or triple the world price. Japan, industrialized and affluent, is able to afford such a policy, but it would be unwise for agrarian countries to emulate that nation. It would seem far better for the less industrialized rice-growing countries to maintain a modest support price and to stimulate rice production by investments in irrigation systems, storage facilities, fertilizer plants, and other requirements for a sound and permanent rice industry.

A few countries in Asia have developed large export markets for their agricultural and industrial products and have a favorable balance-of-payments situation. Malaysia, Korea, and Taiwan are prime examples. All three, however, are attempting either to achieve or to maintain self-sufficiency in rice. The wisdom of these goals may be questioned. In Malaysia, for instance, the government can purchase high quality rice from Thailand at a lower price than it can buy it domestically. Social as well as economic factors govern the adoption of rice policies, and undoubtedly those countries feel there is a political advantage to striving for self-sufficiency.

The Latin American countries should carefully consider their pricing policies for rice. In recent years several have set too high a support price, and the governments have had to sell rice at a loss. Unlike Asia, Latin America still has a relatively low per capita consumption of rice. Thus there is a decided limit to the amount of rice that can be sold domestically. Undoubtedly, the world demand for rice will eventually be great enough so that Latin American countries might become important suppliers. In the meantime, they would be wise to limit their rice production to domestic needs and to whatever export markets they can develop. Moreover, they may find it advantageous to promote the consumption of rice in their own countries, for there are many flat-lying, poorly drained areas in Latin America that are ideally suited to rice growing and that have limited use for other purposes.

Export Marketing: Problems in Marketing Surplus Rice

Many Latin American and African nations have a major comparative advantage in producing rice. They have large areas of land suitable for expanded rice production, with an adequate water supply and a favorable climate. With the implementation of additional research and rural education programs, with an infrastructure to supply inputs and marketing outlets, and with the development of irrigation facilities, the prospect is not only for rice self-sufficiency but, also, for extra amounts of that commodity for export. Understandably, some nations hesitate to take the needed steps, because they fear creating serious marketing problems through excess production. This applies not only to countries with a still underdeveloped rice-growing potential but, also, to those on the verge of self-sufficiency or now producing small surpluses of the grain.

The correct policy decisions revolve around the possible export markets for rice in the foreseeable future; the prospective prices; the grain qualities and types demanded by foreign markets; the kind of approaches made in developing export markets; and the problems of improving the internal purchasing, storing, drying, milling, packaging, and shipping facilities that are needed for a rice export industry.

World Rice Markets

In recent years, between 7 and 8 million tons of milled rice have moved annually from surplus to deficit countries. This is only about 4 percent of total yearly world rice production, but this percentage has held relatively constant in years of low prices as well as in years of high prices. Although some importing countries have been decreasing their per capita consumption of rice, the demand for it in other importing areas has increased. Many of the developing nations are important rice consumers and, as their incomes increase, so does their demand for rice. For example, in Africa, though some countries

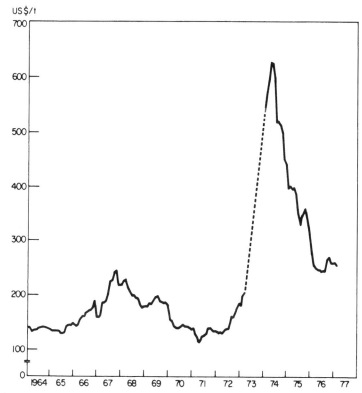

Figure 12. Export price of rice (Thai, 5 percent broken) FOB Bangkok, 1964-77 (from April to December 1973 no prices were quoted). (Source: IRRI)

will increase their rice production, the expected expansion in their populations will cause demand for imported rice to continue. Such requirements will more than offset the reduced consumption in other countries.

In addition, some exporting nations will gradually be forced to reduce their exports of rice as their domestic needs grow. The export demand for rice likely will expand to at least 10 million tons in the 1980s. That 2-million-ton increase over present rice exports would provide an outlet for countries seeking world markets for their surplus rice production.

World Rice Prices

From 1971 through 1977, world export prices for reasonably good quality, 5 percent broken grain content, long-grain rice varied from a low of US$125 per ton in 1971 to a high of US$630 per ton in 1974 (Figure 12). The low point was maintained for only 2 months and the high point for 1 month. On account of good harvests from 1974 onwards, the price of rice had decreased to about US$250 per ton by 1977. Prices for other qualities, including long-grain rice with higher breakage, medium-grain rice, and short-grain rice, as well as for parboiled rice of different qualities, followed the same trends but at US$25 to US$75 per ton less.

World rice prices will continue to fluctuate as they have in the past. With the steady to increasing demand, an average export price slightly higher than that of recent years should be expected—but not so high as that reached in 1974. In general, any nation that can produce rice at an export price of US$250 per ton for high-quality, long-grain rice, or US$175 to US$200 per ton for lower quality types, should be able to compete reasonably well in the world export market. These estimates are based on 1976-78 input cost levels; should inflation continue, the maximum cost levels will need to be adjusted accordingly. Countries whose rice production costs exceed these maximum levels should not strive to produce for export under current price conditions.

Rice Qualities and Types Demanded by Export Markets

Although high-quality, well-milled, long-grain rice with 5 percent or less broken grain commands the highest prices in export markets, there is demand for a wide variety of other rice types and qualities. All three basic types—long-, medium-, and short-grain rice—are in demand. Lots with up to 35 percent broken grain can be sold. High-quality parboiled rice is in great demand in some parts of the world. Medium-quality parboiled rice is marketed in still other regions. Lots of 100 percent broken grain are imported by some countries for blending with local whole grain rice; and a large market exists for broken grain rice in the brewing industry, where it is the preferred ingredient in the production of high-quality beer. Some low-quality rice, especially lots with discolored grains, numerous brokens, insect contamination, undesirable odors, and excessive foreign materials, although salable in the country of production, are not acceptable in export markets.

Developing Export Markets

Rice export marketing requires the establishment of contacts abroad in principal importing centers, the development of lines of credit, the generation of confidence that the deliveries will be of the quality specified, and the organization of the lines of communication that are so important to international trade. The major rice-exporting nations have built this commercial network over a long period. For a country with a sudden sizable surplus to be marketed abroad, this gradual approach may not be possible. Instead it may pay to hire an international grain marketing company to handle the exports.

There are several international grain marketing firms that have numerous contacts with all principal rice-importing countries and with important transportation, storage, financial, and international sales agencies. These private businesses operate to make a profit, and their charges may be large. Thus, it is wise to make the selection through a competitive bidding process. The firms are professionals in the field of grain marketing. For a new rice-exporting nation their charges are likely to be more than worth the investment during the first years of a rice export program.

Internal Improvements Needed for a Rice Export Industry

As a country raises its rice production and moves toward production of a surplus for export, a number of bottlenecks in the marketing system (second generation problems) may begin to develop. In accelerated rice programs in most countries, production increases occur irregularly, with some farming areas expanding more rapidly than others. The first sign of success in the production program, and of the developing second generation problems, is the appearance of these scattered "islands of surplus." Generally, marketing agencies, storage facilities, and established channels of trade are inadequate to handle surpluses. As production continues to expand, additional islands of surplus develop and the distress market situation is intensified. As a result, paddy prices to farmers may decline suddenly and drastically, and affected growers may discontinue their efforts to produce more rice.

To avoid this situation, major adjustments in domestic rice marketing programs must be made early in an overall rice expansion program. In most countries, second generation rice marketing problems require adjustments in the price support program and expansion of rice drying and storage facilities. In many countries, milling facilities must be improved to handle expanded production for export markets efficiently. Some nations will also have to establish standard weights and measures and set up training programs to develop competent personnel in the marketing chain for grading, operating storage facilities, milling, transportation, and merchandising.

The private sector can provide many of the facilities required for both local and export marketing. Where a free economy is in operation, the government should encourage the participation of commercial interests and thereby save public funds for other development activities.

5
Some Successful Rice Production Programs

A number of countries considered to be less developed have nevertheless been able to obtain high average rice yields (4.0 to 5.5 t/ha) and to improve remarkably their rice production over what it was 20 to 30 years ago. Some have made rapid progress just during the past decade and are continuing to do so, although their yields do not yet approach those of Japan or the United States. Furthermore, in certain other countries there are localities, districts, or regions that have mounted extremely successful accelerated rice production programs even though gains in rice yields for the nation as a whole have not been notable.

Such advances are not random developments; they are the result of deliberate and careful planning followed by action. There is no comprehensive blueprint for agricultural development that is appropriate for all environments and all cultures. Certain common elements, however, are applicable to any rice production program in any country that is plagued by low yields and a rice deficit.

The countries (or localities or regions) cited here are simply examples; other high-yield models could have been chosen. The only requirements for inclusion (other than having a successful program) are that the region or country be classed as "less developed" and that adequate information about its rice production be available.

The selected group—which includes Taiwan, South Korea, the Philippines, and Colombia—exemplifies, in addition, the varying conditions in which progress is being made in

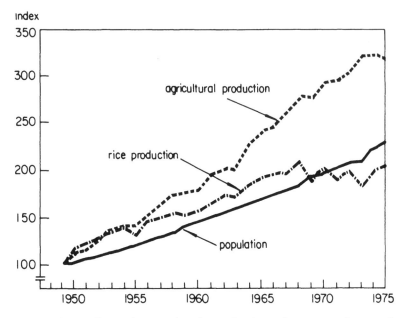

Figure 13. Indices of agricultural production, rice production, and population in Taiwan from 1950 to 1975. (Source: K. T. Li, 1977, "Strategy for Rice Production in Taiwan," *Proceedings: Food Crisis Workshop* (Mimeo) Ramon Magsaysay Award Foundation, Manila)

increasing rice production by raising yields rather than by enlarging the area planted to rice. Taiwan is an example of a situation where yields have increased steadily since World War II in spite of the small size of farms and the vicissitudes of a monsoon climate. South Korea was chosen because of its recent success in adding an extra ton per hectare to its rice yields (which were already the second highest in Asia). The Philippines is pointed out because it is rather typical of the countries of Southeast Asia that traditionally have had, and still have, low rice yields. That country, however, has launched an intensive effort that has had a significant impact on national yields, though leaving considerable room for improvement. Colombia was selected because it is a well-documented example from Latin America of the successful introduction of modern Asian rice varieties on large irrigated farms. The Colombian experience can serve as a valuable model for other countries in Central and South America that wish to expand their rice production.

Figure 14.   Yield of rice (paddy) in Taiwan, 1938 to 1975 (3-year moving average). (Source: Taiwan Provincial Food Bureau)

Taiwan

The average yield of rice (paddy, or rough rice) in Taiwan is about 4.5 t/ha. In 1952 it was only 2.5 t/ha. The area devoted to rice, as well as to agriculture as a whole, has decreased since 1962 because land formerly used for crop production is needed for housing, factories, and highways. Nevertheless, rice production kept pace with population growth until 1967 (Figure 13). Since that time, total rice production has leveled off while population has continued to grow at about 1.9 percent annually.

The yield of rice from 1938 to 1975 is plotted in Figure 14. To moderate year-to-year variations due to weather, a 3-year moving average is used. The year 1938 was chosen as the starting point, because it represents the highest yield obtained up to that time while Taiwan was under Japanese rule. Most of Taiwan's fertilizer then came from Japan. When the latter became involved in the war with China, and later in World War II, its shipments of chemical fertilizer to Taiwan were seriously curtailed. This caused the yield of rice to drop from 2.8 t/ha to about 1.8 t/ha between 1938 and 1946. From 1946 to 1971, yields increased steadily and then leveled off at about 4.5 t/ha. This rise of more than 2 t/ha in less than 3 decades is a remarkable achievement.

Taiwan uses its rice land intensively. About 60 percent of the land produces two crops a year and essentially all of it grows at least one crop annually, with a multiple cropping index in many sections of 2 to 3. The island's soils are not highly fertile, and with such intensive cropping it is unlikely that rice yields will be much higher in future, although modest increases are still possible.

Irrigation

Taiwan was a colony of Japan from 1895 to 1945. Under Japanese administration many irrigation systems were built, the main objective being to increase Taiwan's capacity to produce rice for export to Japan. In 1895, rice was grown on 200,000 hectares, about half of which were irrigated. By 1941 the irrigated rice land had expanded by a factor of five to 547,000 hectares, of which 462,000 hectares were under the jurisdiction of farmer-controlled irrigation associations (the remainder being irrigated by private individuals or groups). Since the end of the war and the end of Japanese rule, several large reservoirs have been constructed for supplying both irrigation water and hydroelectric power. The latest, completed in 1973, made it possible to add 50,000 hectares of land for the production of lowland rice.

Today, 98 percent of Taiwan's 540,000 hectares of rice is irrigated. Because the subtropical climate permits the growing of two rice crops on three-fifths of this land, the total area planted to rice in a single year is about 790,000 hectares. Taiwan, like much of monsoon Asia, has heavy rainfall (1500 to 2000 millimeters annually), but most of it occurs from June to October. Without irrigation, nothing can be grown from November to May (except in the most northerly portion of the island). In spite of the high rainfall, dry spells occur frequently during the monsoon season that would seriously depress yields were it not for the availability of irrigation water.

Varietal Improvement

When the Japanese first occupied Taiwan, the local rice varieties were indicas, originally brought to the island by farmers who had migrated from the Chinese mainland. These indica varieties, which were termed "native" to distinguish them from the japonica varieties introduced from Japan, were planted over most of the rice land of Taiwan into the 1920s.

Although Japanese varieties were introduced into Taiwan as early as 1912, they did not yield as well as the native varieties, particularly in the southern part of the island and in the lowlands in general. Therefore, beginning about 1926, a rather thorough rice breeding and testing program was started for developing japonica varieties suitable for the more tropical climate of Taiwan. Between 1931 and 1943 researchers developed a series of japonica varieties called *ponlais* ("heavenly rice") that yielded as well as or better than the local varieties and provided the grain quality that appealed to the Japanese. Furthermore, the photoperiod insensitivity of the ponlais permitted two crops a year to be grown.

Among the varieties created during this period was Taichung 65, which had a yielding capacity under ideal conditions of over 9 t/ha. It matured in only 120 days and thus fitted into the multiple cropping and rotational systems of Taiwan. Taichung 65 not only was widely grown until about 1959, but it served as a parent in many crosses made during the 1950s. Other ponlai varieties followed Taichung 65, the more outstanding of which were Chianan 8, Chianung 242, Kaohsiung 53, and Tainan 1, 3, and 5.

The most popular variety grown on Taiwan in the 1970s is Tainan 5. It has a high yielding capacity, considerable disease resistance, and a height of only 100 to 110 centimeters. It matures in 120 days in the first crop (March-April to late July) and in 95 days in the second crop (July-August to October-November). Today over 40 percent of Taiwan's rice area is planted to Tainan 5.

In the 1950s, the government of Taiwan supported a limited breeding program to improve the native (indica) rice varieties. A number of superior varieties came out of this program, the most outstanding of which was Taichung Native 1. This variety made a great contribution to rice breeding throughout the world in the 1960s and 1970s, frequently being used as a source of dwarfness, of high harvest index, and of heavy-tillering capacity. By 1965, 79,000 hectares of Taichung Native 1 were planted on Taiwan in the first crop. Since then its popularity has decreased, but as late as 1970 it was second only to Tainan 5 in area planted.

Without question Taiwan's rice breeding program has paid off richly. Although to show their true yield potential these modern varieties require good water control and the abundant use of fertilizer and other inputs, without their adoption average yields in Taiwan would be considerably lower than they are today.

Fertilizers

In 1952, Taiwan used 362,000 tons of fertilizer on its rice crop. By 1975 fertilizer use had almost doubled, amounting to 665,000 tons. Thus in 1975 the average farmer applied 144 kg/ha of nitrogen to each crop of rice and about one-fourth as much phosphorus and potassium. However, in recent years the amounts of phosphorus and potassium applied to rice have increased faster than that of nitrogen, indicating that the soils are becoming depleted of those elements. Because of heavy fertilizer use, rice fields in Taiwan rarely show signs of nutrient deficiency.

Taiwan's practice of using generous amounts of fertilizer is backed by numerous fertilizer trials that have shown there is a profitable response to nitrogen and increasingly so to phosphorus and potassium. Certainly Taiwan's rice yields could not have been realized, nor continuously maintained, without heavy fertilizer applications.

Pesticides

Pesticide use in Taiwan has increased remarkably. Expenditures for pesticides (and a few other materials) in 1975 were US$33 million, a thousandfold increase since 1952 even after discounting inflation. Until recently, the expenditures were mostly for insecticides. Lately, however, as industry has absorbed more and more of the rural labor force, farmers have bought larger and larger amounts of herbicides. As recently as 1972, for example, 79 percent of the expenditures in this category were for insecticides and only 15 percent for herbicides. The remaining 6 percent covered other materials (not including fertilizer). By 1975 expenditures for herbicides had increased to 30 percent, and expenditures for insecticides had fallen to 63 percent. Of course, the actual amount of insecticides and herbicides bought during that 3-year period increased.

The total expenditures for insecticides, herbicides, and incidental materials almost equal the amount spent for fertilizers, US$37 million. Thus, Taiwan's farmers, who cultivate 790,000 hectares of rice, are spending about US$43 per hectare for pesticides and US$47 per hectare for fertilizer.

Less affluent and less industrialized countries should not attempt to emulate Taiwan in the use of pesticides. Every effort should be made to use varietal resistance and integrated pest control to keep down insect infestation. Furthermore, in labor-surplus economies, particularly with irrigated transplanted rice, hand weeding should continue to be used instead of herbicides.

Production Incentives

It had been government policy for Taiwan to produce enough rice to meet domestic demand. As the population and industrialization increased, however, holding to that goal became more difficult and required various changes in the methods of implementing the policy, which had been established in 1949. Authorities who have written the story of Taiwan's strategy for rice production divide it into two periods: 1949-69, and the years since 1969.

During the first period, the aim was to provide low-cost food for the labor force involved in industrial development. Thus performance was not measured in terms of farmer income but rather in terms of rice yield. The government controlled the price of rice but kept it at a low level relative to the cost of other goods, with the objective of transferring agricultural capital formation to industrial development.

In the first period of development, fertilizer supply was inadequate for agriculture as a whole, and the government gave a high priority to rice production. Rice farmers were allowed enough fertilizer provided that they gave paddy in exchange for it. This so-called rice-fertilizer barter system, along with further collections in settlement of land taxes and land price repayments, allowed the government to purchase 26 percent of the rice output. Because only 55 percent of the crop was marketed, it was relatively easy for the government to stabilize prices and to provide cheap rice for industrial laborers, for the urban population in general, and for the military. Large buffer stocks were built up and exports were possible.

Land reform occurred during the first period: land rent reduction was ordered in 1949, and the transfer of land ownership from landlord to tenant took place in 1953. By contributing to the stability of rural society and to better income distribution, land reform had a favorable impact on rice production. The government also spent large amounts of money for rural infrastructure such as irrigation systems, rural electrification, road building, and marketing facilities, all of which contributed to rice production.

By 1965, industrialization had grown tremendously, prices had risen, and many farmers were seeking alternative sources of income. In other words, rice production was becoming unprofitable. Furthermore, due to population increase, the domestic demand for rice equalled the local supply. A change in government policy clearly was needed if Taiwan was to continue to produce enough rice to feed its people. Therefore, in 1969 a new agricultural policy was announced. It was designed to reduce the cost of rice production and to increase the income of the rice farmer, thereby inducing him not only to continue growing rice but to maintain high yields. This new policy included the following elements. (1) subsidizing prices for fertilizer, pesticides, and farm machinery; (2) strengthening farmers' organizations and their services; (3) providing adequate long-term loans for rice farmers; and (4) increasing support for rice research to raise the level of technology available to the farmer.

Announcement of the new policies was followed by action. In 1970, for example, the price of urea was reduced from US$135 per ton to US$116, and ammonium sulfate dropped from US$82 per ton to US$72. In 1971, prices were reduced again—urea to US$100 per ton and ammonium sulfate to US$65 per ton. And in 1972 the rice-fertilizer barter system was abolished.

Finally, in 1974, the government declared a guaranteed minimum price for rice, based on the estimated cost of production plus a 20 percent profit. Through this program, the administration collected approximately one-third of the off-farm rice stocks and then set its price to consumers. Although the actual government price for rice had risen gradually since 1949, in terms of purchasing power it had remained fairly constant. The change in policy in 1974 caused the price index for rice to climb markedly. By 1975 rice prices were 2.4 times their 1971 level. With the price of fertilizer and of pesticides subsidized, Taiwan's farmers now can plant rice with the assurance that if their yields are high they will make a good profit.

Organization of the Farmers

Taiwan has three systems of farmer organization that apply to rice production: farmers' associations, irrigation associations, and joint farming operations and land consolidation.

Farmers' associations. The farmers' associations of Taiwan are controlled by farmers but supervised by the government. Their primary function is to serve the economic needs of the farming population. Although for supervisory and organizational purposes there are one provincial and 20 county-level or city-level associations, the main functional units that serve the farmer directly are at the township level. Within each of these township associations (which, in 1976, numbered 273) are a few small agricultural units located in the villages within each township. These small units serve as a bridge between the township association and its member farmers, distributing useful information and providing voting locations at times of officer election in the town associations. There are over 4500 such small units in Taiwan. One person in each household engaged in agriculture can become a voting member. Today about 95 percent of all the farming households of Taiwan belong to farmers' associations.

Each township association has, on the average, six extension agents who advise farmers on the technological advances. Much of the extension education is carried on through farming study groups in the villages. One survey indicated that two-thirds of the increase in yield of agricultural crops should be attributed to improved technology that has been spread throughout the region by over 1600 full-time agricultural extension agents employed by the farmers' associations.

The farmers' associations make production loans to farmers and accept savings deposits from association members. The profit made by lending money at a somewhat higher rate of interest than that paid for savings deposits provides the funds for operating the credit program. At the end of 1975, the total credit extended to farmers amounted to US$359 million, of which three-fourths was for farming use and the rest for nonfarming purposes. At the same time, the farmers' associations held US$471 million in savings deposits. About 90 percent of Taiwan's farmers obtain credit from the farmers' associations.

Distribution of chemical fertilizers and the storing and milling of rice are highly important functions of the farmers' associations. The Food Bureau of the government makes contracts with the farmers' associations to distribute fertilizer and also arranges for the associations to store and mill about one-third of Taiwan's rice crop (the portion usually purchased by the government). To be able to provide such services, the farmers' associations own 1600 rice warehouses, which can store about 850,000 tons of paddy rice, and 800 fertilizer warehouses with a capacity of about 400,000 tons. In addition, the associations own 400 rice mills with an average daily capacity for each mill of 10,500 tons. Beside fertilizer, the associations sell to farmers seeds, agricultural machinery, pesticides, and other products.

In summary, the farmers' associations contribute immensely to the agricultural development of Taiwan. They provide farmers with a source of agricultural supplies, credit, and technical advice. Through membership in the associations farmers gain a sense of participation. Furthermore, the support of government leaders has given the associations stability and the confidence of the farming population.

Irrigation associations. The irrigation associations of Taiwan are self-governing corporate bodies organized by farmers to administer irrigation systems and to construct new ones—but under close government supervision. There are 16 irrigation associations, each with four divisions: engineering, management, finance, and administration.

In 1975, the irrigation network of Taiwan consisted of 190 main canals, 817 laterals, and 444 sub-laterals, with a total length of over 36,000 kilometers. In years of normal rainfall, the systems can supply water continuously to all farmers in the command areas. However, in recent years, rotational irrigation has been introduced to save water in times of drought. In this system, each irrigation area is divided into sections that take turns in using the irrigation water at 3-day to 5-day intervals. Water savings of up to 25 percent have been reported from the use of rotational irrigation.

Major irrigation projects are subsidized by the government, but from 30 to 50 percent of the costs are provided by the irrigation associations. In turn, the associations get their funds from membership dues. The fees are determined by land productivity and by the source and costs of water. The charges that farmers pay vary from as little as US$1.50 per hectare annually to as high as US$25.00 per hectare.

Taiwan's system of managing its irrigation works through special associations has resulted in the best operated and maintained irrigation system in Asia (with the possible exception of Japan). This, in turn, has had a profound impact on rice yields.

Joint farming operations and land consolidation. Two relatively new activities, joint farming operations and land consolidation, are not normally classified as "farmers' organizations." Both are designed to increase the efficiency of farming, particularly with regard to mechanization and labor efficiency.

Since 1968 there has been a strong effort to help farmers combine in the farming of 20-hectare blocks. The land that makes up a block is operated as one farm, with the labor (usually 30 to 40 families) and farm equipment already contained in the area.

Consolidation of fragmented small landholdings in Taiwan

began in 1959. The procedure is to rearrange the small plots in irrigated areas into rectangular shapes of larger size. By 1971, over 260,000 hectares of farm land had been consolidated. Joint farming operations for rice, at least, have been most successful on these consolidated areas.

As wage rates increased in relation to capital costs, Taiwan's rice industry has become more mechanized. From 1966 to 1976, for instance, the number of power tillers rose from 9300 to about 45,000. Land consolidation and joint farming operations have further stimulated mechanization. In the last several years, sales of rice transplanting machines, combines, and four-wheeled tractors have risen by 30 to 100 percent annually.

Off-farm Opportunities

Although possibly not directly affecting the yield of rice, the abundant opportunities for off-farm employment in Taiwan contribute greatly to its economic prosperity and to the general well-being of both urban and rural (and thus rice-farming) populations.

There is much manufacturing in the cities, but small industries are scattered throughout the countryside. The types of industry found in rural areas tend to be agricultural businesses, such as food processing plants, starch factories, and plants manufacturing farm tools and equipment. In addition, there are small nonagricultural establishments making such products as plastics, textiles, and parts for the electronics industry.

Many a farm family owning less than a hectare of land is able to maintain a relatively high standard of living because several of its members are employed in nearby factories. The high degree of literacy in Taiwan is possible partially because children can be spared from the farm to attend school. This is in great contrast to the situation in poorer countries where school-age children are needed on the farm to help eke out a living.

Taiwan's foreign exchange position is excellent. Its industrial and agricultural prosperity is reflected in a per capita income in 1977 of just over US$1000 annually, which—with the exception of Japan and Singapore—is the highest in Asia. Unlike many less affluent countries in South and Southeast Asia, Taiwan has reached a stage of development that allows it to subsidize the rice industry and thus transfer income to the rural areas.

Summary

Taiwan has included in its strategy for rice production all the important elements, the most crucial of which are irrigation, heavy investment in rice research, the wide use of modern rice varieties, high fertilizer and pesticide application, and a well-organized system of farmers' associations to supply extension services and credit. To this list must be added access to reliable markets, guaranteed minimum prices for rice, subsidized prices for inputs, and abundant off-farm employment opportunities. Finally, mention must be made of the industriousness of Taiwan's farmers and their willingness to accept and comply with various governmental regulations designed to boost agricultural development throughout the land.

Countries with little industrialization and scarce foreign exchange resources will have difficulty in adopting Taiwan's plan in its entirety. Nevertheless, Taiwan stands out as an example of what can be done with small farms in the tropics and subtropics; and many parts of its program are applicable in the less developed countries.

South Korea

Rice yields in South Korea have climbed steadily during the past several decades (Figure 15). There have been wide variations from year to year, often caused by drought. As irrigation facilities have been improved, however, yields have become more stable.

South Korea has long had a nationwide program to increase rice production, and by 1970 it reached a national average yield of 4.6 t/ha. In spite of this high yield, however, the nation still had to purchase rice abroad to satisfy its domestic needs. From

Yield of paddy (t/ha)

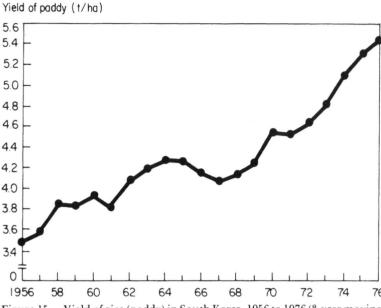

Figure 15.   Yield of rice (paddy) in South Korea, 1956 to 1976 (3-year moving average). (Source: FAO)

1969 through 1972, South Korea annually imported 500,000 to 900,000 tons of milled rice. During this period the government greatly intensified its effort to produce more rice. As all the suitable rice land was already in use for that crop (the harvested rice area in South Korea has remained between 1.1 and 1.3 million hectares since 1957), the only course open to agricultural planners was to raise yields. Between 1972 and 1976, South Korea was able to add an extra ton per hectare to its average rice yield. This was a remarkable achievement considering the already high level of national yields.

Before turning to the factors responsible for this outstanding success, the future of the upward trend might be pondered. Although the steepness of the yield curve during the past 4 to 5 years would indicate that yields will continue to rise during the next few years, experience in other countries strongly suggests that the ceiling has almost been reached. No country, including Japan, has been able to maintain average yields above 6 t/ha; and the FAO estimate of South Korea's average rice yield in 1976 is 5.9 t/ha. Even though the nation has reduced its annual population growth rate from 3.0 percent in 1960 to 1.7 percent, population increase will probably make South Korea a rice-importing nation again by 1985.

Fertilizer used (ooot)

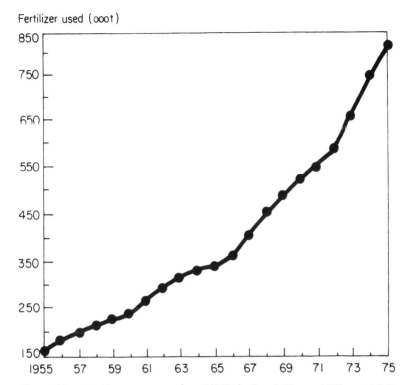

Figure 16.   Fertilizer consumption (NPK) in South Korea, 1955 to 1975 (3-year moving average). (Source: FAO)

Fertilizers

The soils of South Korea are derived largely from granite and therefore tend to be sandy with low levels of native fertility. Consequently, good soil management, including heavy applications of fertilizer, is essential for high yield.

No data for South Korea are available on the amount of fertilizer used on rice alone. Nevertheless, as a high priority crop, it is likely to receive a large share of the fertilizer allocations. Fertilizer use has risen rapidly during the past 20 to 25 years, particularly since 1969 (Figure 16). In that year, the country consumed 478,000 tons of fertilizer (expressed as nutrients NPK). By 1974, fertilizer use had climbed to 757,000 tons. This is an unusual increase for a 5-year period. Before 1957, South Korea used less than 200,000 tons of nutrients annually.

Today South Korea manufactures all of its chemical fertilizer. Its current high rice yields could not have been achieved without the ample investment made in the construction of chemical fertilizer plants and in the importation of the raw materials needed for their production.

Irrigation

The government of South Korea has spent large sums on irrigation. In 1966-68, for example, over US$60 million was allocated to irrigation projects, which was more than one-third of the entire investment in agriculture during those years. In 1975 alone, the expenditures for irrigation projects were over US$30 million. However, because the investment in agriculture was much greater by then, that figure represented only 15 percent of funds spent for agricultural development.

Although 85 percent of the nation's rice crop is now irrigated, 30 to 40 percent of it is poorly irrigated, and there is an inadequate supply of water during drought periods. Apparently, the 15 percent of South Korea's rice land that is now rainfed cannot be economically irrigated. Nevertheless, the country can wisely invest additional funds in improving those areas now classed as "irrigated," but which need an assured water supply throughout the growing season.

Varietal Improvement

South Korea has long had a rice breeding and varietal testing program. In 1962, the system of agricultural research and extension was reorganized and consolidated into the Office of Rural Development, which is directly under the Minister of Agriculture and Forestry. The Office of Rural Development now has responsibility for both research and extension activities, including community development projects related to agriculture. The reorganization stimulated agricultural research and extension, in which programs rice continued to receive top priority.

In the mid-1960s, among the 24 leading rice varieties being grown in South Korea, half were developed by Korean plant breeders, and the remainder were imported directly from Japan, a nation that always has been a leader in rice research. Because South Korea's climate is similar to that of central and northern Japan, many of the better varieties did as well in South Korea as in Japan. However, in spite of the advances made in their own program and of the adaptability of Japanese varieties, Korean farmers were plagued with outbreaks of the rice blast disease and the stripe virus and with lodging when high amounts of fertilizer were applied.

In an effort to remove those yield constraints, a cooperative research program between South Korea and the International Rice Research Institute was started in 1965. The principal cooperators in South Korea were the Office of Rural Development and the College of Agriculture of Seoul National University. Both organizations sent research scholars to IRRI for training in plant breeding and in other disciplines.

Research scholars who worked in rice breeding under the direction of IRRI scientists made a number of japonica-indica crosses in an attempt to develop varieties that had more disease resistance than the japonica varieties then being grown in South Korea and that also possessed the desirable plant type features of IR8. By including a japonica variety they hoped to retain the low amylose content of the grain desired by the Korean consumer.

Among the first crosses made at IRRI and tested in South Korea was one between Yukara (a Japanese variety) and Taichung Native 1 (a short-statured indica variety from Taiwan), from which an F_1 progeny was then crossed with IR8. The most promising selection from that triple cross was designated as IR667-98. Although this new selection did not have sufficient cold tolerance or an ideal grain shape for the Korean market, in widespread yield trials in 1970 it outyielded the standard varieties then grown locally by 30 to 40 percent. Its high yield potential can be attributed largely to its resistance to the rice blast disease and the stripe virus, to its early maturity, to its lodging resistance, and to its profuse tillering capacity. Seed of IR667-98 was multiplied in South Korea during the summer and at IRRI in the winter. In 1972 Korea gave this selection the varietal name "Tongil." That same year Korean farmers grew 187,000 hectares of the variety.

Tongil needed further improvement. While it was being multiplied and distributed to farmers, the rice breeding program continued at an intensive pace. As a result, several "Tongil-type" varieties were developed. Among the better ones are Yushin, Milyang 21, and Milyang 23. These varieties have greater cold tolerance and better grain quality than Tongil without any lowering of yield potential. IRRI has continued to cooperate in both the breeding and the seed multiplication programs, but most of the work and the costs have been the responsibility of the Koreans.

In 1977, the improved japonica-indica varieties of the Tongil type were grown on 57 percent of South Korea's rice land. Government planners expect that by 1980 about 90 percent of the nation's rice farmers will be planting these improved varieties. Although other factors have contributed to the rapid increase in rice yields in South Korea since 1972, without the greater disease resistance and the improved plant type that came out of the rice breeding program, the nation could not have attained yields that are now the highest in Asia.

Cultural Practices

South Korea's abundant rice harvests have resulted to a significant degree from excellent cultural practices. Aside from the improvements in irrigation systems and the increase in fertilizer use, Korean farmers have applied minor elements on about 37 percent of the rice-producing land where one or more trace elements are known to be deficient. The use of silica has increased most notably.

The nation's farmers diligently control weeds by hand or by applying chemical herbicides. Recently, there has been a tremendous increase in the use of vinyl-covered seedbeds to permit early planting, so the farmers can harvest their rice 2 weeks earlier and follow it immediately with a crop of barley. The use of vinyl covering in seedbeds rose from 4 percent in 1971 to 65 percent in 1976. Pest control has been greatly improved by the use of warning systems and the application of pesticides over wide areas as needed.

Price Incentives

South Korea has a guaranteed minimum price for rice and also subsidizes inputs. In the mid-1970s, the country's procurement price for rice was set as high as US$270 per ton. At the same time the government subsidized fertilizer at 44 percent of the normal retail price. According to the Asian Development Bank, a kilogram of nitrogen cost 1.42 times as much as a kilogram of paddy in 1976. In contrast, in Thailand, where fertilizer is not subsidized and the purchase price for rice is low (usually about half that of South Korea), nitrogen cost 4.08 times as much as paddy in 1976. South Korea, like Taiwan, has shown a high industrial growth rate in recent years and thus can afford to offer excellent price incentives to rice farmers. For many rice-growing countries, a pricing policy that provides greater incentives than does Thailand, but with support prices lower than those in Taiwan and South Korea, would be appropriate. Nevertheless, the high prices paid for rice in South Korea have permitted farmers to purchase inputs readily and to improve their living standard. In the 1970s Korean farm families greatly increased their savings, as compared with the amounts they put away in the 1960s.

The New Village Movement

In 1970 the South Korean government launched a massive community development program called the *Saemaul Undong* —the New Village Movement. This nationwide program was initiated by the president of the country and has had his strong support ever since. Although the program involves much more than increased rice production, it has provided information, inspiration, and technical guidance to the rice farmer and has much improved the rice-growing environment through the rural development projects included in the plan.

The New Village Movement is a joint effort by the government and the people to attain a higher standard of living in South Korea's 34,660 villages by attacking the causes of rural poverty through a massive self-help program with strong support from the central government. Substantial funding was contributed by the government, much of it in the form of cement and structural steel for road building, for the improvement of irrigation and drainage ditches, for the construction of wells where appropriate, and for other modernization projects, such as public buildings. However, the value of the completed projects has been much greater than the resources contributed by the government (Table 10). In

TABLE 10. GOVERNMENT CONTRIBUTIONS AND THE TOTAL
VALUE OF THE COMPLETED PROJECTS IN THE
NEW VILLAGE MOVEMENT IN SOUTH KOREA
FROM 1971 TO 1976

Year	Value of government contributions (US$ millions)	Estimated value of the completed projects (US$ millions)
1971	8.8	24.0
1972	6.6	63.0
1973	43.0	196.0
1974	61.6	265.6
1975	330.0	591.8
1976	330.0	645.0

other words, the self-help component was truly significant.

In 1974 all projects connected with increased food production (including such projects as irrigation and flood control that, in part at least, had been handled previously by separate agencies) were encompassed in the New Village Movement. This partly accounts for the fivefold increase in government contributions in 1975 as compared with those for 1974.

Another factor in the success of rice growing among "Saemaul" farmers was the increase in cooperative farming units from 22,000 in 1972 to about 52,000 in 1977. These units allowed farmers to share equipment and labor and, by working larger areas of land, to increase the efficiency of their operations. This movement was similar to the joint farming in Taiwan.

Again through the New Village Movement, but administered by the Office of Rural Development, extension activities were widely expanded from 1972 to 1976. Near the start of the intensified program, 1800 new extension specialists were added, and by 1976, 7500 were working in the rice-growing villages. The inauguration of special training courses for extension workers enabled them to advise farmers properly in the growing of the Tongil-type rice varieties and in basic objectives of the New Village Movement. These extension

TABLE 11. AVERAGE INCOMES OF URBAN AND RURAL
FAMILIES IN SOUTH KOREA FROM 1970 TO 1975

Year	Annual income of urban families (US$)	Annual income of rural families (US$)
1970	762	516
1971	904	712
1972	1034	859
1973	1100	961
1974	1289	1349
1975	1718	1745

people, plus others who were not rice specialists, ran massive winter training courses for farmers. In 1976 alone, 2.8 million Korean farmers attended the short courses.

These activities are only a small part of the New Village Movement which included all important rural activities, such as wheat and barley production, horticulture, livestock farming, forestry, and fisheries, as well as activities in private industry and some urban areas.

The standard of living of the rural people in South Korea has improved much, as indicated by the increased use of television sets, motorcycles, refrigerators, home water systems, and methane gas installations for cooking. The impact on rural life can best be seen, however, by comparing average family incomes in rural and urban areas (Table 11). Although rural incomes were considerably lower than urban incomes in 1970, by 1974 they were higher.

Summary

South Korea, like Taiwan, is an example of a country that has mounted an extremely successful national program to achieve self-sufficiency in rice production. The average yield had reached the respectable figure of 4.6 t/ha by 1969, indicating that even then many of the requirements of high yield and a thriving rice industry had been met. These prerequisites included reasonably good irrigation facilities, a prosperous fertilizer industry, good roads and communication facilities, adequate access to dependable markets, a well-organized rural credit system, a completed land reform program, and satisfactory price incentives.

Notwithstanding the high yields already being obtained in the late 1960s, South Korea found that its domestic requirements for rice exceeded what it was producing on about 1.2 million hectares of land. Therefore, it decided to make an all-out effort to increase yields still further. It is much more difficult to boost yields from 4.6 t/ha to 5.6 t/ha than it is to go, for example, from 2 t/ha to 3 t/ha. Thus South Korea's feat of adding an extra ton to national yields in a 5-year period can be considered a unique accomplishment.

The nation achieved self-sufficiency in rice by 1976. The most recent spurt in yield was brought about mainly by the breeding of a new set of modern rice varieties that were short and heavy tillering, with strong sturdy straw that resisted lodging at high fertility levels and that were resistant to South Korea's two most serious rice diseases, rice blast and stripe virus. In addition, cultural practices were bettered, irrigation facilities were renovated and enlarged, and the extension services were expanded and improved.

It is doubtful that all of this would have been accomplished had there not been a strong national will, at the popular as well as the official level, to make such progress and had that will not been implemented through the New Village Movement. Incorporated in this movement were massive indoctrination and intensive extension education. All the forces of the countryside were mobilized. Villages, and individual farmers within them, that did not measure up to expectations were singled out for special attention.

As in Taiwan, farmers responded well to government directives to a degree that might be difficult to attain in countries where the population has not yet been asked to accept such regimentation. Nevertheless, South Korea remains an excellent example of a country that, despite small farms, relatively infertile soils, and a limited land area suitable for rice production, has been able to boost its yields to exceptionally high levels.

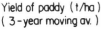

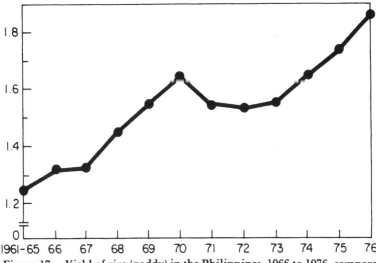

Figure 17. Yield of rice (paddy) in the Philippines, 1966 to 1976, compared with the 1961-65 average (3-year moving average). (Source: FAO)

The Philippines

The Philippines never has had a high national average rice yield. Before 1966 this could be attributed chiefly to the growing of tall, lodging-susceptible rice varieties and to the fact that less than 15 percent of the rice was fully irrigated, most of it being vulnerable to drought or flood depending on the vagaries (including typhoon incidence) of the monsoon rains. From 1961 to 1965—before the advent of the Green Revolution —the average yield of paddy in the Philippines was only 1.2 t/ha (Figure 17). Among Asian rice-growing countries only Laos and Kampuchea had lower yields during that period.

The first spurt in Philippine rice yields came in 1969 and 1970, largely as a result of the introduction of the modern rice varieties developed at the International Rice Research Institute. Between 1966 and 1970, the percentage of the rice-growing area planted to the improved varieties rose from essentially zero to over 50 percent—the fastest adoption of the new varieties in all of South and Southeast Asia.

Yields slumped in 1971 and 1972 because of severe drought, typhoon damage, and a widespread outbreak of the tungro virus disease. During 1973-77, yields climbed slowly but steadily. There appear to be two principal reasons for the gains. One is that the government launched an intensive rice production program, known as Masagana 99. The other is that the country was continually expanding its irrigation system for rice.

The Masagana 99 Program

Since the rice crop in the Philippines in 1971 and 1972 was severely affected by unfavorable weather and, particularly in 1971, by an outbreak of the tungro disease, the country was badly in need of more rice. The secretary of agriculture was aware of the potential for increased production, and in 1972, with the assistance of both IRRI and the Philippine agencies concerned with rice research and production, he launched a pilot project to test known varieties and techniques on entire farms, rather than on plots within farms. In addition, a start was made in organizing leaders at barrio (village), municipal, and provincial levels and in improving the infrastructure, including rural credit.

The new pilot extension program involved the cooperation of the National Food and Agricultural Council, the Bureau of Agricultural Extension, provincial organizations in Central Luzon, and IRRI. Pilot barrios within municipalities were selected for their "progressiveness." One trained technician was assigned to five or six barrios, in each of which the better farmers were selected to try a package of practices on their land. The rural banks were allowed to provide noncollateral loans to farmers who either leased land or were strictly tenants. During 1972 this pilot scheme covered about 2600 hectares. The average yield on the closely supervised farms exceeded 4 t/ha, whereas surrounding farms outside the pilot extension scheme were obtaining no more than 2 to 3 t/ha.

As a result of the pilot extension program, and of the promising additional applied research results obtained in 1972, the Philippine government decided to launch a massive nationwide rice production program. Accordingly, in 1973 President Marcos announced an all-out effort to bring rice production in the Philippines to at least the level of self-sufficiency and labeled the program Masagana 99. Masagana is a Tagalog word meaning "bountiful," and the figure 99 quantified the goal of the program to bring yields up to 99 cavans (44 kilograms each) per hectare. In other words, the objective was to raise yields to about 4.4 t/ha on farms that entered the program.

Principal elements. The four elements that formed the backbone of the program were credit, the transfer of the new technology (including a massive publicity program), price support for rice, and the provision of low-cost fertilizer.

From past experience in the Philippines, it was evident that the majority of farmers could not purchase the inputs needed to increase rice production unless they had access to credit. Furthermore, most of them were either tenants or leaseholders and did not have collateral to secure loans. In 1973-74, the government set up an expanded and revolutionary credit system involving 420 rural banks, 102 branches of the Philippine National Bank, and 25 offices of the Agricultural Credit Administration. The government assured the various credit organizations that it would guarantee 85 percent of all loans if the agencies would make production loans to farmers who had no collateral but who were enrolled in the Masagana 99 program. As a result, the banks no longer were reluctant to make loans to small farmers.

Every effort was made to simplify and speed up loan applications. Bank representatives and field technicians processed loan applications in the field, obviating the necessity for farmers to spend time going to the bank to apply. The Philippine National Bank, for instance, purchased 100 jeeps in order to deliver money to the farmers' doorsteps. The farmers had help from the field men (and women) in drawing up farm plans and sensible budgets that would be acceptable to bankers. Often farmers were organized into groups (called *seldas*) of 5 to 15 who were jointly responsible for one another's loans. By the end of the first full year of operation (1974), Masagana 99 farmers had been granted 643,000 loans, amounting to US$80 million. In addition, 257,000 farmers participating in the Masagana 99 program got along without borrowing. Thus 900,000 farmers had joined the program.

Credit was not unlimited. A maximum of about $100 per hectare was available to each farmer who qualified for a loan. Part of each loan was provided in cash to cover labor costs, and the balance was given out in purchase orders for fertilizer, pesticides, etc. The merchants who sold those products to the farmers were able to redeem the purchase orders for cash at the banks.

Just as important as credit were technical advice and guidance so that the farmer could wisely invest the funds he had borrowed. This service was accomplished through an extension program backed by 3200 well-trained technicians, all graduates of Philippine agricultural colleges, who were deployed to the villages.

The extension program was an integral part of Masagana 99 and had three main features. First, widespread trials of new varieties and management practices were carried out on farmers' fields. Second, the field technicians visited the Masagana 99 farmers frequently, advising them on the more promising varieties, the appropriate fertilizer application for their particular soil type, the use of pesticides, and so on. They also assisted them in making loan applications and in turn helped the banks in making the collections when the loans matured. Third, as an additional aid in transferring the modern technology to the small farmer, a greatly expanded program to disseminate information through the mass media was undertaken.

A private advertising agency donated its services to produce the largest radio broadcasting campaign that the Philippines had ever known (a survey having shown that three out of every four Filipino farmers had a radio). The company produced a series of down-to-earth spot announcements, radio skits, and musical jingles in the six principal languages of the Philippines. Over 50 1-hour radio broadcasts were given daily, usually by the field technicians, providing timely information on the management of the rice crop.

Supplementing the radio broadcasts, hundreds of thousands of pocket-size Masagana 99 primers, leaflets, and brochures (some even in comic style) were printed in the six languages. As an added publicity feature, membership flags were distributed to permit Masagana 99 farmers to advertise their cooperation with the national program to make the Philippines self-sufficient in rice again.

The government support price for paddy in the Philippines increased steadily, rising from US$97 per ton in 1972 to US$170 in 1976, an increase of 75 percent. However, during the same period, the consumer price index rose by about 70 percent. Thus, in real terms, the government purchasing price changed little.

Although the Philippines was not able to subsidize fertilizer to the extent that South Korea did, in the first year of the Masagana 99 program it subsidized urea by 21 percent and in 1975 increased the subsidy to 25 percent. The high cost of fertilizer in 1974-75 lowered its use by farmers, but the great reduction in world fertilizer prices in the next two years again made it profitable for farmers to use fertilizer. The leaders of the program considered that the 25 percent subsidy of fertilizer costs was significant to the early success of the effort.

The impact on yield. Although national average rice yields have not yet reached 2 t/ha, the leaders of Masagana 99 reported that in 1974 and 1975 rainy season yields on 610,000 hectares of irrigated rice averaged 3.4 t/ha and on about 300,000 hectares of rainfed paddy, 2.8 t/ha; in the dry season 590,000 hectares of fully irrigated rice yielded 3.85 t/ha. Yields among Masagana farmers were about the same in 1976 and 1977 as in 1974 and 1975.

Without doubt, the Masagana 99 program contributed significantly to the increase in rice yield and in total production, especially in 1974 and 1975. The reports issued on Masagana 99 state that yield increases by the 900,000 participating farmers ranged from 0.4 to 1.2 t/ha, depending on the level of their former yields and the extent to which they had adopted modern practices. However, the fact that national average yields continued to rise in 1975, 1976, and 1977, whereas the average yields on Masagana 99 farms remained at about the same level, indicates that the effort has not been completely successful and, furthermore, that factors outside the program were contributing to the increase in national average yields.

One disappointment of the Masagana 99 program has been the marked decrease in loan repayments since the start of the program. The Asian Development Bank reports that in 1973-74 the repayment rate was 91 percent. In 1974-75 it decreased to around 76 percent. By 1975-76 it had gone down to 35 percent. Recent indications are that there has been no improvement in the rate of loan repayment. Unless specific action is taken, there is danger that the movement, as such, will stagnate at its present level.

Expansion in Rice Irrigation

Unfortunately, accurate data have not been published on the amount of rice land in the Philippines that has been converted from rainfed to irrigated. In 1977 the secretary of agriculture stated that the Philippines was bringing 100,000 hectares of rice a year under irrigation. If we assume that irrigated rice produces 1 t/ha more crop than rainfed rice in the wet season and that the irrigated land is double cropped with an average yield of 3.5 t/ha in the dry season, then 100,000 hectares of newly irrigated rice land should add 450,000 tons of rice to the annual harvest in the Philippines, which is currently about 5.5 million tons.

Summary

The rice production program in the Philippines is an example of a national effort, in a typical rice-growing country of Southeast Asia, to attain self-sufficiency in rice. Since the nation imported no rice in 1975 and 1976 and actually exported a small surplus in 1977, its immediate goal was reached. Although the weather was better than average during those 3 years, it was not progressively better each year.

The disturbing fact in the Philippines is that despite Masagana 99, the presence of IRRI, and the strong national effort to increase rice production, national average yields are still less than 2 t/ha. According to studies by IRRI scientists and by other agencies, the apparent reasons for the low yields are inadequate irrigation, inappropriate fertilizer use, and poor pest and weed control—in that order of importance. There may be both economic and social causes for these deficiencies in technology. The fact remains, however, that until the rice fields are well managed, yields will continue to be below the potential. Another obvious reason why Philippine rice yields are low is that of the 3.6 million hectares devoted to rice, about 500,000 are planted to upland rice, the yields of which are extremely low. Furthermore, consideration must be given to the fact that the Philippines has the highest incidence of typhoons of any Asian country.

Although, statistically speaking, 40 percent of the Philippine rice crop is irrigated, many of the irrigation canals and ditches are poorly maintained, and water deliveries are not properly managed. Too often, farmers located at some distance from the source of irrigation water are not supplied at critical times. Furthermore, many irrigation systems provide water for only short periods beyond the duration of the rainy season. So for many farmers, irrigation serves, at best, only to supplement the rains during periods of drought or to extend slightly the growing season for a single crop of rice. Because the Philippines is continually improving and enlarging its irrigation systems for rice, however, yields undoubtedly will continue to rise during the years ahead.

Such a forecast for progress presupposes that there will be no slackening of government support for improving and expanding irrigation systems, for agricultural research and extension, for a workable credit system, and for a guaranteed minimum price for rice. The goal of the Masagana 99 program to achieve a paddy yield of 4.4 t/ha is a reasonable one for fully irrigated land, and a national average yield approaching 4.0 t/ha can be attained if there is a united and sustained effort by the government and the farmer to bring it about. With the present high population pressures in the Philippines, there appears to be no danger of overproduction; a good domestic market for rice is assured indefinitely.

Colombia

For the two decades ending in about 1965, rice yields in Colombia remained slightly less than 2 t/ha. But between 1966 and 1975, the national average yield of rice more than doubled

Yield of paddy (t/ha)

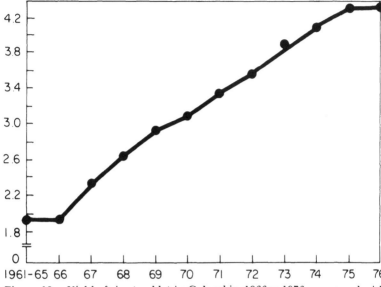

Figure 18.   Yield of rice (paddy) in Colombia, 1966 to 1976, compared with the 1961-65 average (3-year moving average). (Source: FAO)

(Figure 18). The most important causes of this dramatic increase were (1) the adoption of modern, stiff-strawed, disease-resistant rice varieties as replacements for the taller U.S. varieties and (2) an increase in the proportion of rice land under irrigation.

Introduction and Breeding of Modern Varieties

Up to 1957, the dominant rice variety in Colombia was Bluebonnet 50, a rather tall U.S. plant that had excellent grain quality. That year there was a widespread outbreak of a virus disease known as hoja blanca, which is transmitted by a rice planthopper *(Sogatodes oryzicola)*. Bluebonnet 50 proved to be susceptible, and Colombian scientists sought resistant varieties. In 1961, another U.S. variety, Gulfrose, which showed some resistance to the vector of the hoja blanca, was released to rice growers. In 1963 a cross between Bluebonnet 50 and Palmira 105 (a local selection), named Napal, proved to be resistant to the virus disease and was released to farmers. However, it, too, soon succumbed to the disease.

Next, in 1965, Tapuripa, a variety from Surinam, was

released and became popular with the rice growers. By 1968, over 40 percent of the irrigated rice land in Colombia was planted to that variety. Tapuripa's advantages were that it was vigorous and sturdy and yielded at least 1 t/ha more than Bluebonnet 50.

In 1067, CIAT (the Spanish acronym for the International Center for Tropical Agriculture) was established in Cali, Colombia. It immediately set up a rice research program with cooperation from IRRI in the Philippines. The program was conducted in conjunction with the Colombian Department of Agriculture (ICA) and the Colombian National Federation of Rice Growers (FEDEARROZ). CIAT introduced a number of IRRI's best short-strawed varieties and genetic lines. Several of those introductions were resistant to the planthopper that spreads the hoja blanca virus and gave high yields under Colombian conditions. The first semidwarf varieties to be widely grown in Colombia were IR8, IR22, and CICA 4. The first two were IRRI-named varieties; CICA 4 was a Colombian selection from an IRRI genetic line. Later (1975), CICA 6 was released.

In 1970, 36 percent of the irrigated rice land in Colombia was still planted to Bluebonnet 50, with 26 percent sown to Tapuripa and 29 percent to IR8. By 1976 the amount of Bluebonnet 50 had become negligible, and 9 percent of the irrigated land was planted to IR8, 33 percent to IR22, 34 percent to CICA 4, and 22 percent to CICA 6. Thus, modern, short-statured varieties by then occupied essentially the entire irrigated rice area in Colombia.

No doubt the fact that FEDEARROZ was keeping abreast of recent developments in rice technology aided greatly in the rapid adoption of the new varieties. The basic reason for such acceptance was that the new varieties not only had strong resistance to the planthopper vector of the hoja blanca virus, but they outyielded the best varieties formerly available by about 2 t/ha when properly managed.

By 1976 Colombian rice researchers had developed two other varieties, CICA 7 and CICA 9, each of them with a higher yield potential, as well as greater disease resistance, than previously used varieties and with grain quality superior to that of IR8.

CICA 9 is somewhat taller than the other Colombian varieties, has great vigor, and is heavy tillering. Therefore it may prove to be suitable for the more favorable upland sites, as well as for irrigated and rainfed paddy fields. In addition, it is resistant to many races of the rice blast disease that occur in Colombia and in other Latin American countries. In short, it promises to make a sizable contribution to higher rice yields not only in Latin America but in similar environments in other parts of the world. Undoubtedly CICA 9 will replace many of the earlier varieties in Colombia and elsewhere.

Although the increased use of pesticides, fungicides, and improved cultural practices had a definite influence on Colombia's yields of irrigated rice, the main reason why yields jumped from 3 t/ha to over 5 t/ha in less than 10 years is that Bluebonnet 50 was replaced by modern short, fertilizer-responsive, disease-resistant varieties.

Irrigation

As Colombian rice growers were adopting new varieties, another change was taking place that had a profound impact on national average rice yields: the proportion of rice that was irrigated increased and the area devoted to upland rice correspondingly decreased (Table 12). In 1966, before the accelerated rice production program got under way, upland rice occupied two-thirds of the total rice area and irrigated rice, one-third. By 1975, upland rice had shrunk to only a quarter of the rice area. In that same period, the total land area planted to rice had increased by only 5 percent, yet annual rice production in Colombia rose from 680,000 to 1,622,000 tons.

Irrigated rice in Colombia has consistently yielded more than twice as much as upland rice (Figure 19), even when Bluebonnet 50 was the predominant variety. Therefore, as the percentage of irrigated land increased, average national yields likewise increased (Figure 20). The high correlation (0.98) between yields and amount of irrigation should in no way detract from the contribution of modern varieties and other inputs. It simply shows that good water control is essential for the full expression of varieties and of such inputs as fertilizers, pesticides, and herbicides.

TABLE 12. AREA, PRODUCTION, AND YIELD OF RICE IN COLOMBIA FROM 1966 TO 1976 (BY SECTORS), AND THE PERCENTAGE OF THE RICE AREA UNDER IRRIGATION

Year	Area (thousand hectares)		Production (thousand tons)		Yield t/ha			Irrigated (% of total)
	Upland rice	Irrigated rice	Upland rice	Irrigated rice	Upland rice	Irrigated rice	National average	
1966	235	114	339	341	1.44	3.00	1.94	32.6
1967	180	110	280	381	1.55	3.47	2.28	37.8
1968	150	127	251	535	1.67	4.22	2.84	45.8
1969	135	116	220	474	1.64	4.09	2.77	46.2
1970	121	112	198	554	1.64	4.94	3.22	48.0
1971	109	144	174	731	1.59	5.06	3.57	56.8
1972	103	171	161	883	1.56	5.17	3.81	62.3
1973	99	192	155	1021	1.56	5.32	4.04	66.0
1974	96	273	150	1420	1.57	5.20	4.26	74.0
1975	96	286	151	1471	1.60	5.10	4.25	74.9
1976	95	261	148	1333	1.50	5.10	4.16	73.2

Data from Colombian publications that gave the source as statistics released by FEDEARROZ.

Yield of paddy (t/ha)

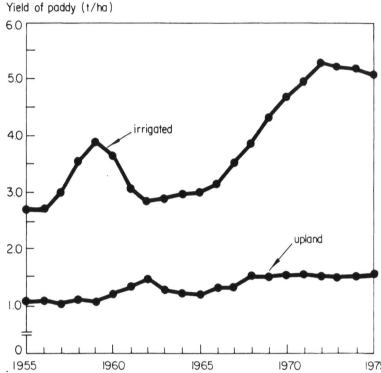

Figure 19. Average paddy yields in Colombia under irrigated and under upland conditions, 1955 to 1975 (3-year moving average).

The government investment in irrigation is estimated to be about US$2000 per hectare. Between 1966 and 1975, the irrigated rice area enlarged by 160,000 hectares. This represents an investment of US$319 million in irrigation. The additional rice produced in 1975 alone (as compared with 1966) was 952,100 tons. The cost of obtaining this extra rice must be divided between the research that went into developing the new varieties and other techniques and the investment in irrigation.

From experimental data available in Colombia, it is impossible to calculate absolutely the impact of irrigation in contrast to that of varieties. However, if the assumption is made that the varieties did not change but that the areas devoted to the two systems of rice culture would change as they did between 1966 and 1975, the average national rice yield in 1975 would be 2.6 t/ha. If, on the other hand, the varieties changed as they did and yields increased particularly on the

Grain yield (t/ha)

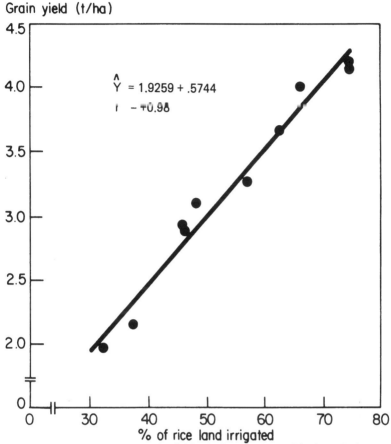

Figure 20. Average national yield of rice in Colombia in relation to percentage of rice land irrigated.

irrigated land, but there was no change in the proportions of irrigated and upland rice, the average yield would be 2.8 t/ha. Neither of these assumed situations is realistic, as the availability of the modern, high-yielding varieties stimulated the use of irrigation and discouraged the production of rice under upland conditions, where the modern varieties could not express their superior yield potential.

The only conclusion that can be reached is that both varieties and irrigation played highly significant roles in increasing rice production in Colombia. If neither the varieties nor the proportion of irrigated land had changed, it is rather likely that yields would have remained near the 2 t/ha that existed for the decade before 1966. As it was, by changing both factors national average yields increased to the remarkable level of 4.3 t/ha by 1974.

The National Federation of Rice Growers

An important contribution to the rapid increase in the average yield of rice in Colombia was made by an efficient organization dedicated solely to serving the rice grower. Similar to the farmers' associations of Taiwan, FEDEARROZ (National Federation of Rice Growers) is an organization that, although originally formed by the growers themselves, now receives its principal financial support from the government. FEDEARROZ was established in 1947 to promote the political and economic interests of rice growers, but its function as a technical agency did not start until 1963. In that year, a law was passed that levies a tax of one Colombian cent on each kilogram of rice sold by the rice grower. This amount is deducted from the purchase price to the rice buyer, who, however, is required to pay it later to the government. The law allows FEDEARROZ to handle the funds collected and to use them for the support of rice research, for regional variety testing, for publishing technical bulletins, for running training courses for field agronomists, and for covering other expenses for its technical division. In addition to its technical services, FEDEARROZ produces and handles certified rice seed and sells fertilizer, pesticides, herbicides, and farm machinery at reasonable prices. The organization is described as a nonprofit agency that provides services and supplies to farmers. Its well-trained technical advisory staff keeps abreast of the latest developments in rice research in Colombia and elsewhere and brings the newest techniques and information to farmers.

Although the rice farmer actually pays the cost of the benefits derived from FEDEARROZ, farmers recognize that it is a well-run, highly efficient organization, designed to serve their interests whether political, economic, or technical.

The Use of Inputs

The Colombian rice farmer, on irrigated land at least,

applied high amounts of fertilizer even before the advent of the modern varieties. It appears that the amount of fertilizer used on rice in Colombia did not increase as the modern varieties were introduced. Nevertheless, there is no indication that the new varieties did not receive adequate amounts of fertilizer.

There is evidence that, from 1968 until 1974, rice growers' use of insecticides and fungicides increased markedly. This accounts, in part, for the rise in yields on irrigated rice land even before the full impact of the modern varieties was felt. Rice is mostly direct-seeded in Colombia and chemical herbicides are widely employed. The total amount of herbicides used climbed particularly during the expansion of irrigated rice land between 1971 and 1974.

Other Considerations

In contrast with a number of Asian countries where price supports, subsidized inputs, and improved credit facilities played important roles in raising yields and production, in Colombia those factors did not change greatly as rice yields expanded. From 1966 to 1975 there was a good domestic market for rice, and all production was consumed in Colombia (14 percent of the calories in the Colombian diet are derived from rice). Since the support price for rice usually was at or below the world price, the government seldom bought any. The abundant supply of rice following the adoption of the improved varieties pushed prices down and provided a cheap source of food for the low-income, urban population. This can be construed as a social benefit of the accelerated rice production program.

In 1976 the rice situation in Colombia changed somewhat. The area of irrigated rice decreased by 25,400 hectares and upland rice by 400 hectares. Yield levels of both irrigated and upland rice were stable, but total production fell by 141,000 tons, 9 percent below the record level of 1975. The basic cause for the reduced rice harvest was economic, not technical. According to authorities in Colombia, the reduction in area and production (which continued in 1977) was due to a combination of high production costs and low farm prices for rice. Farmers switched to other crops, believing that the profits would be greater. The decrease in production forced Colombia to import 33,000 tons of rice to satisfy domestic requirements.

It appears that the Colombian government might find it advisable to increase the support price for rice modestly to guarantee adequate rice for domestic requirements. Furthermore, the results of recent research obtained at CIAT indicate that the Colombian rice grower could reduce production costs by using less fertilizer, by practicing more efficient methods of weed control, and by sowing smaller quantities of seed—without reduction in yield.

There has been no subsidization of inputs in Colombia, except that FEDEARROZ buys fertilizer, pesticides, herbicides, and farm machinery, and sells them to farmers at the lowest price possible.

Although there was no apparent increase in the amount of credit extended to rice farmers as yields went up, there also was no credit decrease. The Caja Agraria and the Fondo Financiero Agrario, the two important credit agencies in Colombia, made loans to rice growers from 1968 to 1974 averaging US$30 to US$40 per hectare annually. Most individual loans were larger than that, because some rice producers did not need to borrow money.

Transferability of the Colombian Experience

Rice growing in Colombia is unlike rice growing on the small farms of Asia. Over 60 percent of Colombia's rice crop is produced on farms larger than 50 hectares. Many farms are 100 to 500 hectares, and a few are over 2500 hectares. Most are highly mechanized. In fact, the techniques of growing rice in Colombia (and in much of the rest of South America) are more similar to those used in the United States than in Asia. Therefore, the applicability of the Colombian experience is mostly confined to those regions where landholdings are large and where heavy machinery is available for land preparation and crop harvesting.

In South America (and in some parts of Central America and Africa) there are vast areas of low-lying flat land, particularly in the larger river basins, that are particularly well adapted to rice growing. Many parts of these areas are so poorly drained that no commercial crop other than rice can be grown; they remain largely unused today. They could be converted at relatively small cost into rainfed rice land by the construction of rather simple dikes or bunds to control the water level. With greater expenditures, the same areas can be irrigated for year-round rice production. On much of this potential rice land the varieties and methods used in Colombia would be appropriate.

The lesson to be learned from the Colombian experience is that with good water control, with modern varieties, and with the ample use of fertilizers, insecticides, fungicides (in South America fungicides are commonly used for the control of the rice blast disease) and herbicides, yields of over 5 t/ha can be obtained.

Latin American countries are in an excellent position to take advantage of Colombia's successful rice production effort. They can send both research and extension staff members to CIAT for training. They can arrange for delegations of large rice farmers to observe rice farming in Colombia first hand, and they can directly use some of the varieties developed by CIAT and by the Colombian Ministry of Agriculture.

There seems to be little question that rice production in Latin America will continue to increase and that, concurrently, per capita consumption of rice will rise (provided the price remains low in relation to that of other staple food crops). Supporting good research programs directed toward finding ways of producing rice more efficiently is an important means by which governments can make rice growing more profitable. There is abundant evidence that rice research, both national and international, has paid handsome returns during the past two decades.

Other Countries That Have Made Rapid Progress

In selecting the four countries as examples of those that have made rapid progress in increasing rice yields and total production during the past decade or so, an attempt was made to use well-documented cases in widely varying environments. There are more than four successful examples, of course. Mention, however brief, should be made of several other countries that have shown at least a 25 percent increase in average yields of rice from 1961-65 to 1971-75.

In Pakistan, rice yields during that 10-year period jumped 63 percent, from 1.42 to 2.31 t/ha. This increase was brought about primarily by substituting modern varieties such as IR8 for traditional low-yielding varieties. Nevertheless, yields are still much lower than they should be. All of the rice is irrigated, and because of the arid climate solar radiation is high. Yields on experimental fields often are between 7 and 10 t/ha, showing that the potential exists. Without doubt, if the effort is made, Pakistan can have an average national rice yield of between 5 and 6 t/ha.

The principal reasons for the low national yield appear to be poor water control, high salinity and alkalinity in some areas, and the continued production of Basmati rice for export in spite of its low yielding capacity (though new varieties are being created that have the excellent grain quality of the Basmati rices but that are short and stiff-strawed and respond well to heavy applications of fertilizer).

In Indonesia, yields of paddy rice increased about 25 percent between 1961-65 and 1971-75, to about 2.5 t/ha. Indonesia made a strong effort to increase rice production through a national program similar to the Masagana 99 project in the Philippines. An analysis by the Asian Development Bank attributes the increases in rice yield in Indonesia to two main factors. One is the rising proportion of rice land under irrigation and under rainfed paddy conditions relative to the area devoted to upland rice. The other is the widening use of chemical fertilizer. Moreover, the amount of credit extended to farmers has expanded pronouncedly. Unfortunately, as was true in the Philippines, the repayment of loans to farmers made under crop intensification programs in Indonesia has dropped—from 91 to 95 percent in 1971-73 to only 44 percent in 1975.

India achieved only a 16 percent gain in rice yield between 1961-65 and 1971-75. Nevertheless, certain states and regions in India made remarkable advances in yield. In the northwestern region consisting of Delhi, Himachal Pradesh, and the Punjab, rice yields increased 84 percent to 2.8 t/ha. Similarly, in the more northern Jammu and Kashmir region, yields rose 82 percent to 2.7 t/ha. Less outstanding but above average gains were obtained in the southern region comprising Andhra Pradesh, Tamil Nadu, Karnataka, and Kerala, where yields (already at high levels in the early 1960s) increased 25 percent to 2.6 t/ha. On the lower end of the scale was that portion of eastern India containing the states of Orissa, West Bengal, and Bihar, where yields went from 1.4 t/ha in 1961-65 to only 1.5 t/ha in 1971-75, an increase of 7 percent.

The causes for these wide variations in yield gains have not been thoroughly studied. Still, certain differences among the areas mentioned are obvious. The two regions in northwestern India where exceptionally rapid progress was made grow mostly irrigated rice and have much higher levels of solar radiation than do the less productive areas. Furthermore, the incidence of insect and disease attack is much lower there than in more humid areas.

The states in south India cited as making above average progress had initiated, early on, sizable research and production programs with resulting yield gains that, in the first half of the decade under consideration, exceeded the rate of rice improvement in the dryer, irrigated states of the northwest.

The regions of eastern India showing little advance in rice yields contain extensive areas of rainfed rice and are subject to both drought and floods. Moreover, partly because of the uncertainties of the weather and the consequent economic risk involved, considerably less fertilizer is used, and cultural practices in general are less intensive than in the high yield regions. Unfortunately, appropriate modern varieties that are superior to local varieties under such unfavorable environmental conditions have yet to be released.

Laos and the Ivory Coast showed large yield increases during the decade under consideration. However, the yields in both countries have not only remained at low levels but have shown no further increases since 1973.

The Soviet Union has made remarkable progress in raising rice yields and total production. Between the 1961-65 and 1971-75 periods, yields increased by more than 50 percent, area almost tripled, and total rice production more than quadrupled. Russia irrigates all of its rice crop and has developed superior high-yielding varieties in a well-supported rice research program.

Comparing the Programs

The three countries that have national average rice yields in excess of 4 t/ha (South Korea, Taiwan, and Colombia) have had the following conditions in common: (1) a high percentage of the rice crop is produced under irrigation; (2) the irrigation systems are well managed and good water control exists throughout the growing season; (3) the use of fertilizer and pesticides is adequate; (4) good weed control is practiced; (5) modern, short-statured varieties with a high yield potential are widely grown; (6) the government supports an adequate rice research program; (7) sufficient credit is available to farmers; (8) a well-trained extension staff is at hand to assist farmers in obtaining higher yields; and (9) there is at least one farmers' organization with the sole purpose of providing the services that farmers need.

The Philippines, in strong contrast to the other three countries used here as examples, as of 1977 had yet to obtain an average national yield in excess of 2 t/ha. Yet there is ample evidence of the possibility of attaining yields of over 4 t/ha on its irrigated land and on rainfed lowlands where adequate water control is possible. The same is true for lowland rice in all of South and Southeast Asia, in Latin America, and in Africa.

Undoubtedly, lowland rice yields will continue to rise, particularly in Asia where little land exists for expanding the rice-growing area. The yield increase probably will come about through better water control, higher fertilizer applications, and the development and use of varieties that have high levels of tolerance to variations in water level, to adverse soil conditions, and to insect and disease attack.

One pattern that is consistent wherever rice is grown is that national average rice yields are low if a high percentage of the rice crop is grown under upland conditions. In Southeast Asia, Latin America, and Africa no country has been able to obtain average upland rice yields of 2 t/ha, and in many regions yields

TABLE 13. PROPORTION OF RICE AREA UNDER LOWLAND CONDITIONS, NATIONAL RICE YIELDS, AND YIELDS OF LOWLAND AND UPLAND RICE IN 10 COUNTRIES

Country	Year of data	Proportion of total rice area planted to lowland rice (%)	Yield (t/ha)			Advantage of lowland rice relative to upland rice (%)
			National average	Lowland rice	Upland rice	
Indonesia	1972	86	2.26	2.46	1.23	100
Korea	1974	98	5.13	5.17	2.60	99
Philippines	1974	87	1.60	1.71	0.85	101
Taiwan	1973	99	4.08	4.09	2.33	76
Ivory Coast	1963	14	0.89	1.35	0.82	65
Madagascar	1961	84	1.82	2.05	0.57	240
Brazil	1976	22	1.33	3.35	1.20	180
Colombia	1976	74	4.42	5.10	1.50	225
Peru	1976	75	4.78	5.55	1.74	219
Venezuela	1974	28	2.52	4.67	1.68	178

Note: Lowland rice includes all irrigated rice and rainfed rice that is bunded. Upland rice is direct-seeded in dry soil in fields that are not bunded and requires frequent rains for its moisture supply.

are below 1.5 t/ha. Furthermore, in most nations upland rice yields are only one-third to one-half those of lowland rice (Table 13).

Certainly, as population pressures mount and new land suitable for rice growing is fully used, more and more land will be irrigated, and in the tropics and subtropics it will be double cropped to rice. Furthermore, it is likely that upland rice gradually will be replaced by other crops that have greater drought resistance.

6
Promising Rice Research

International agricultural research centers and national programs in many rice-growing countries are conducting well-rounded research efforts covering all important aspects of rice production. This chapter singles out a few lines of such investigations that show unusual promise of making, within the next decade, a significant contribution toward the removal of serious constraints to high rice yields on farmers' fields.

The purpose is not only to point out the real hope that exists for the further weakening of the barriers to high yield, but more importantly, to identify for research administrators in the rice-producing nations some of the more fruitful research projects that merit concentrated attention. Yield-limiting problems are not simple; only the most thorough and widespread research among the involved countries will resolve them.

The priority given the various lines of research will depend upon the nature and severity of the constraints encountered in a particular environment. Although the areas of research selected for discussion here are unavoidably somewhat arbitrary, they are ones that affect large numbers of rice farmers in many countries.

Varietal Improvement

The yield potential and the disease and insect resistance of the rice plant have improved remarkably since the 1960s. Yet, a number of goals that appeared to be attainable have not been reached. Some of the rice breeding objectives that need to be stressed are early maturity, more stable resistance to insect and disease attack, maintenance of fertilizer responsiveness, and tolerance to drought, to varying water depth, and to adverse soil conditions.

Early Maturity

As population pressures mount, it becomes increasingly critical, especially in the tropics and subtropics, to grow as many crops a year on the same land as possible. Early maturing rice varieties make possible the growing of several rice crops (or rice and other crops) within 12 months. Furthermore, under rainfed conditions, rice varieties that have a short growth duration often yield better than those that have a long growth duration, because they escape the drought that occurs when the rainy season ends early. Although rice breeders have produced early maturing varieties with growth durations of 90 to 110 days, too often such varieties have lacked good plant type, high yield potential, and adequate resistance to insects and diseases.

IRRI plant breeders have developed several genetic lines and several varieties (IR36 is an example) that mature under tropical conditions in about 110 days (from seed to seed), have desirable grain characteristics, and are resistant to most important pests and diseases. However, even earlier maturing varieties are needed. In 1976, IRRI began an intensive rice-breeding project to develop varieties that have exceptionally high vegetative vigor in the early growth stages, mature in 90 to 95 days, and yet have a yield potential of no less than 7 t/ha. It is extremely likely that that program will succeed and, also, that there will emerge from national breeding programs similar varieties with particular adaptability to the regions where they are bred and selected.

More Stable Resistance to Insects and Diseases

Rice breeding work aimed at developing stronger resistance to insect and disease attack has usually involved major single gene resistance (often called vertical resistance). This approach has been effective against the green leafhopper and grassy stunt virus disease but less successful in combatting certain other insects that readily form biotypes (such as brown plant-hoppers) and diseases that develop physiologic races (such as rice blast).

Plant breeders have three principal approaches that might be more successful in obtaining stable resistance than the commonly used method of identifying major genes and transferring them to otherwise suitable varieties through the standard pedigree breeding system. One approach is gene pyramiding. This consists of combining in each variety a complex of both major and minor genes for resistance to a given pest or pathogen.

A second approach is the development of multiline varieties. A series of genetic lines is developed that have similar external characteristics but that differ in their reaction to specific biotypes or physiologic races. The lines are then mixed and planted together. Experience with wheat and other crops has shown that this method can provide protection against losses when new genetic mutants of a particular pest or pathogen appear in an area.

A third approach is to create horizontal (field) resistance by incorporating into varieties many minor genes that provide moderate-level resistance to the insect or disease. In the major or single gene (vertical) resistance now being used so extensively, the continued tolerance of the host plant to the pest or pathogen depends entirely on the stability of a single avirulent gene in the attacking organism. On the other hand, the theory behind minor or polygenic (horizontal) resistance is that so many genes are contributing toward resistance that the pest or pathogen cannot mutate sufficiently to overcome all of them.

There are techniques of screening still to be perfected in the horizontal resistance breeding program. Many procedures being used in conventional pedigree and vertical resistance breeding programs have to be discarded. Nevertheless, results with other crop plants (particularly potatoes in breeding for resistance to late blight) indicate that, if this breeding method were widely used in high-volume polycrossing breeding programs, the stability of resistance to disease and insect attack in rice would be substantially increased.

IR42, a variety that is highly lodging resistant, is still erect a week before harvest, while other promising genetic lines adjacent to it have lodged because of weak stems. Actually, just a few days before harvest the plot of IR42 did start to lodge, indicating that it, too, requires further improvement in straw strength. (Source: IRRI)

Maintenance of Fertilizer Responsiveness

The rice variety IR8 represents an ideal plant type. Its stems are short, thick, and sturdy; hence it does not lodge even when heavy applications of nitrogen fertilizer are made. However, it lacks several important characteristics—particularly, good grain quality and adequate resistance to insects and diseases. Most varieties developed between 1967 and 1977 by IRRI and by several national rice breeding programs reflect efforts to overcome the undesirable traits of IR8. They are far superior in grain quality and in insect and disease resistance. On the other hand, they have weak stems and lodge when grown at high nitrogen levels. On many farms, this characteristic does not depress yields severely, because the farmers apply only modest amounts of nitrogen. But as pressures for greater food production intensify, it becomes increasingly necessary to use larger amounts of fertilizer to approach the true yield potential of the modern rice varieties. If this essential practice is to be followed, rice breeders will need to restore in their varieties the strong sturdy stems they were able to obtain in IR8.

The apparent reason that plant breeders at IRRI and elsewhere have produced so many weak-stemmed rice varieties during the past decade is that a number of selections used as parents (notably TKM6 from India and *O. nivara*) that carry genes for resistance to several important insects and diseases have weak stems. Moreover, it seems likely that there is a genetic linkage between weak stems and resistance to pests and pathogens. If this were not so, breeders seemingly would have retained the character for strong sturdy stems that was present in at least one of the parents of their crosses.

In spite of the complexities encountered in incorporating into one variety resistance to insects and diseases and to lodging, plus satisfactory grain quality, the difficulties probably will soon be overcome. In fact, IR42, an IRRI line named by the Philippine Seed Board in 1977 that has attractive grain quality and good insect and disease defenses, also appears to have more than ordinary resistance to lodging at high nitrogen levels. Lodging resistance will be especially important in varieties for direct-seeding, which may become more common in the years ahead.

Drought Tolerance

In the eleven Asian rice-growing countries listed in Table 3, 34 percent of the rice land is classified as irrigated, 50 percent is devoted to rainfed paddy, 8 percent is used for growing upland rice, and another 8 percent is planted to deep-water rice. Although the amount of irrigated land will continue to increase, with a consequent reduction in the area under rainfed paddy and upland rice, the fact that two-thirds of the rice land in South and Southeast Asia is without a controlled water supply makes it important to continue to develop varieties that can withstand periods of drought without severe reduction in yield. Not only in Asian countries, but also in Brazil and in much of West Africa, drought-tolerant rice varieties are badly needed.

Rice scientists have fully demonstrated that varieties vary in their capacity to endure drought. It seems evident that a deep root system may be the single most important characteristic associated with drought tolerance. Even though rice breeders have long been engaged in developing varieties that can withstand high soil moisture tension, no rice variety yet approaches the drought tolerance of wheat, maize, sorghum, millet, or sweet potatoes. In recent years IRRI has intensified its research on drought tolerance. Rice scientists are attempting to identify plant characteristics that contribute to drought endurance. Hundreds of crosses are made annually, and thousands of the progeny are tested in an effort to combine drought tolerance with other desirable plant qualities. Similar work is going on in West Africa, particularly at IITA in Nigeria and at Bouake in the Ivory Coast.

Great differences exist among rice varieties in the ability to survive under severe moisture stress, but the impact of this quality on yield capacity is not yet clear. It is logical to assume, however, that when the drought-tolerant characters are introduced into otherwise desirable varieties, yields under conditions of moisture stress will be elevated considerably.

Tolerance to Varying Water Depths

In tropical monsoon Asia, many flat low-lying areas are flooded during the height of the rainy season. If water levels reach 15 to 30 centimeters or more, modern varieties are usually unsuitable. In many of the river deltas of India, for instance, farmers grow the new varieties during the dry season but continue to plant the taller traditional ones in the wet season.

Extensive areas in Bangladesh, India, Thailand, and Vietnam and localized areas in all other rice-growing countries flood to depths of 1 to 6 meters. In these areas, only deep-water or floating rice varieties can be grown. About 10 percent of the world's rice land is classified as suitable only for deep-water culture.

An additional 20 percent of the rice-producing area of Asia is subject to water depths of less than 1 meter—but often with flooding during much of the growing season to depths of 30 to 60 centimeters. Some of that same land, during short periods, may be flooded to the extent that young rice plants are completely submerged for a week to 10 days.

It is unlikely that major increases in yield potential of the deep-water varieties that have to endure water depths of over 2 meters will be realized, although there will be some advances tied to such qualities as improved "kneeing" ability and increased survival under deep flooding during early growth stages, as well as to better weed control methods before the flood waters arrive. There is an excellent chance, however, that greatly improved varieties that are tolerant to medium-to-deep water depths and to temporary submergence will be created soon.

Rice breeders have demonstrated that, when a deep-water variety is crossed with a short-statured one, some of the progeny have the capacity to remain short when the water is shallow and to elongate when the water rises. The IR442 lines are examples of this phenomenon. Selections from the IR442 cross have shown wide adaptability, even exhibiting considerable drought tolerance and yielding well under upland conditions. In addition, good harvests have been obtained at water depths up to 60 centimeters. However, none of the lines from this cross, which lack photoperiod sensitivity (desirable to avoid flowering before the flood waters have subsided) and adequate resistance to insect and disease attack, has been named by IRRI.

Rice breeders paid little attention to developing rice varieties for varying water depths until 1975 when IRRI and Thailand began a joint program of breeding and testing in the Philippines and at the Huntra Deep-water Rice Station in Thailand. Furthermore, meetings have been held with representatives from other countries with deep-water rice problems, and an extensive cooperative scheme for breeding and testing rice varieties has been inaugurated. Among the objectives of this cooperative movement is the intention to breed and widely test rice varieties that can survive prolonged periods in water between 30 and 60 centimeters deep. IRRI reported in 1976 that a number of selections from this program consistently yielded over 4 t/ha whether grown in shallow water or in water 50 centimeters deep. In water 100 centimeters deep, however, yield was reduced 28 percent.

Rice scientists are also testing thousands of genetic lines for tolerance to complete submergence during the early stages of growth. Preliminary results indicate that many varieties can withstand complete submergence for a week to 10 days without any significant reduction in yield and that this quality can be combined with tolerance for medium-depth flooding. Fortunately, it is being found that drought tolerance can be combined with adaptability to medium-depth water and to temporary submergence.

In summary, there is genuine promise that a series of disease- and insect-resistant varieties can be developed that combine high grain quality, photoperiod sensitivity, and drought tolerance with the ability to yield well in water as much as 60 centimeters deep, and that can stand total submergence for a week to 10 days during the early growth stages. The successful achievement of such a goal will benefit hundreds of thousands of farmers who are growing lowland, rainfed rice that is subject to the vagaries of the monsoon rains. It is entirely possible (though with no firm evidence yet) that these new varieties, if properly weeded and fertilized, will outyield currently grown rice varieties by 1 to 2 t/ha.

Tolerance to Adverse Soil Conditions

An estimated 40 million hectares in the tropics and subtropics are unsuitable for modern rice varieties because of adverse soil conditions. The most injurious conditions in those areas are salinity, alkalinity, strong acidity, iron toxicity, zinc deficiency, phosphorus deficiency, and large amounts of organic matter (peat soils).

Most detrimental soil conditions can be alleviated by proper soil management, but the treatment is frequently quite expensive. For that reason, IRRI recently greatly expanded its effort to breed and widely test rice varieties that are tolerant to problem soils. In 1976 alone, IRRI scientists screened about 17,500 entries from the germ plasm bank and the breeding and hybridization program. The most encouraging result of the tests was that several named modern varieties and elite genetic lines proved to be tolerant to a number of adverse soil conditions. IR30, IR32, and IR36, for example, showed a salinity tolerance only slightly inferior to that of a well-proven resistant line (IR2153-26-3-5-2). Furthermore, nine elite lines with good plant type and resistance to insect and disease attack had a higher tolerance to alkali soils than did several varieties widely grown in alkali areas. In addition, IR28, IR29, IR30, and IR34 were tolerant to phosphorus deficiency; and IR20 and IR34 showed some tolerance to low zinc levels in the soil.

As in drought tolerance research, many of the tests for resistance to salinity, alkalinity, high acidity, and growth on organic soils reflect survival and vegetative growth, and there is too little information yet available as to the yield levels that can be expected when the genes for resistance to those toxic conditions are incorporated into otherwise superior varieties. Nevertheless, considerable benefit should result from this work in time.

More data are available on the tolerance to low zinc and phosphorus levels. Some modern varieties have yielded as much as 5 t/ha without the addition of zinc or of phosphorus, while less tolerant varieties yielded less than 3 t/ha. No variety, however, can yield well when the level of either zinc or phosphorus is extremely low, unless the soil deficiency is corrected.

Supplying Nitrogen to the Rice Plant

Nitrogen is almost universally deficient in rice-growing soils, but chemical sources of nitrogen are becoming increasingly costly. Three promising avenues of research may reduce farmers' expenditures for fertilizer: improving the efficiency of the utilization of chemical nitrogen by the rice plant, the partial substitution of organic materials for chemical nitrogen, and increasing the biological fixation of atmospheric nitrogen.

Loss of Nitrogen

A rice crop often absorbs no more than 25 to 50 percent of the nitrogen applied to the soil, because much of the nitrogen is lost or unavailable. There are five principal ways in which nitrogen may be lost after it is applied: by ammonium fixation, by direct volatilization of ammonia from the flood water, by immobilization in the soil organic matter (from which, however, some of the nitrogen will be recovered later), by leaching, and by denitrification. A symposium held at IRRI in 1977 made it clear that current quantitative measurements of nitrogen transformation and loss are inadequate. For example, authorities largely agree that most nitrogen is lost through denitrification; yet directly measuring the escape of nitrogen gas from the soil into an atmosphere already laden with the element is highly difficult. Although indirect methods have depicted total losses reasonably well, they have not shown satisfactorily just where the losses occur.

Present knowledge indicates that ammonia gas escapes from the surface of flood water when the pH is high and when nitrogen fertilizer is broadcast onto the surface without mechanical incorporation. Furthermore, deep placement of fertilizer into the reduced zone is known to cut down on denitrification losses. It also has been proved that when a paddy soil dries out, nitrogen depletion through denitrification increases. It is further recognized that when nitrogen fertilizer is applied as a topdressing to a rice crop that has a healthy, active root system, efficiency is high, because the nitrogen is absorbed before it can be transformed or lost. In spite of such practical knowledge, soil scientists remain unable to account for the disappearance of added nitrogen by totaling the values of the various loss components. Furthermore, the size of the experimental error and the great variations in research workers' estimates of nitrogen losses point to the serious need for improvements in research techniques.

The efforts of soil scientists to improve the methodology for measuring the destiny of nitrogen applied to rice should meet with success in the next decade or so. Until that is accomplished, however, progress in improving the efficiency of nitrogen use will be slow.

Nitrogen from Organic Materials

Longtime fertility experiments in Japan, the Philippines, and elsewhere have revealed that from half to two-thirds of chemical nitrogen can be replaced with organic nitrogen from such sources as compost without any reduction in yield. The organic matter source can be compost, oil cake, animal manures, green manures, or rice straw. However, experimental results show that if top yields are to be obtained, some chemical nitrogen is needed either in making compost or when other organic materials are added to the soil. Moreover, materials with a high carbon/nitrogen ratio, such as rice straw, must be applied well ahead of rice planting to avoid tying up too much nitrogen during the early growth stages of the crop and to get rid of toxic organic compounds formed when the straw starts to decompose.

In many of the long-term experiments, 10 to 20 t/ha of compost and 35 to 55 kg/ha of chemical nitrogen were applied to obtain maximum yields. In China it is common to use as much as 75 t/ha of compost. However, the Chinese have found that if maximum yields are to be obtained, such heavy applications of organic matter must be supplemented with chemical nitrogen.

Organic sources of nitrogen increase the organic matter content of the soil, release nitrogen slowly during the entire period of crop growth, and add other plant nutrients, both major and minor. It is probably for good reason that the Japanese farmers who have won the rice yield contests have invariably applied both organic and inorganic fertilizers.

In the developed countries where labor is scarce and expensive, no doubt chemical nitrogen will be used entirely for the foreseeable future. On the other hand, where labor is abundant rice farmers can well use local sources of organic matter, particularly green manures, rice straw and other crop residues, and animal manures, for at least half of the nitrogen required by a high-yielding rice crop.

Soil scientists are studying ways of managing and applying organic matter as a source of nitrogen and other elements. If the results of both past and future studies are brought to the attention of farmers through on-farm trials, it is likely that many of them will change their present practice of burning rice straw simply to get it out of the way. Moreover, improved methods of conserving animal manures are required. Much organic matter is now wasted in rural areas. It is unlikely that the farmers of South and Southeast Asia will go to the same extremes as Chinese farmers in using all sources of organic matter (from night soil to composts made from animal manures and crop residues to city rubbish), but great economies can be achieved by increasing the amounts of organic fertilizer and decreasing the purchase of chemical nitrogen. Agriculturalists must develop sound management systems for maintaining soil fertility by the most economical means in the less developed countries.

Fixation of Atmospheric Nitrogen

Longtime rice fertility trials conducted in both the temperate zone and the tropics have proved that yields between 1.5 and 2.0 t/ha can be maintained year after year on many soils without the addition of nitrogen from either organic or inorganic sources. The amounts of nitrogen removed by the rice crop range from about 35 kg/ha to over 100 kg/ha, with an average around 60 kg/ha. The nitrogen absorbed by an unfertilized rice crop comes largely from the decomposition of soil organic matter and from atmospheric nitrogen that is fixed by soil microorganisms.

Studies on the quantity of nitrogen that is fixed in the root zone in rice paddies and in the paddy waters by blue-green algae and by other nitrogen-fixing microorganisms are conflicting. Usually at least 20 kg/ha is fixed during each crop-growing season, and values three to four times that amount have been occasionally reported. There are doubts surrounding some of the higher values, and it is generally agreed that measurement techniques are still inadequate. Soil incubation studies in the laboratory are often unreliable indicators of what takes place under field conditions. The acetylene reduction method has improved the accuracy of estimates of the amount of atmospheric nitrogen that is fixed, but it has limitations.

There is sound evidence that more nitrogen is fixed in flooded rice soils than under upland conditions. It is likewise known that nitrogen fixation values are higher in the tropics than in northern latitudes. It seems clear, also, that the most abundant microorganisms that fix nitrogen are the blue-green algae and that their capacity to do so in flooded rice fields is considerably enhanced when they are growing in association with the water fern, *Azolla*.

There is great need for expanded research on biological

fixation of atmospheric nitrogen in lowland rice paddies. For example, there appears to be insufficient knowledge about the amount of nitrogen actually fixed not only by the blue-green algae but by other microorganisms. Methods of managing *Azolla*, particularly under tropical conditions, need more study. The impact of the presence of a rice crop on nitrogen fixation is not well understood. These are only a few of the deficiencies in our knowledge of nitrogen fixation in paddy soils.

These problems, and others, have been worked on by scientists with varying and inconclusive results. Soil microbiologists today, however, feel that if nitrogen fixation research is sufficiently intensified throughout the rice-growing world, progress resulting in outstanding benefits to the rice industry is a certainty.

Improved Insect Control at Low Cost

Studies of the yield constraints on farmers' fields have shown that, predictably, the application of insecticides increases yields significantly when insect populations are high. Yet the gains in yield, although substantial, are often unprofitable because of the high cost of the insecticides. Aside from less costly insecticides or other methods of insect control, the solution to the problem may lie in finding ways to increase the effectiveness of insecticides so that good control can be obtained with smaller quantities. As an example, when carbofuran is applied to the root zone, 0.5 kg/ha (active ingredient) gives better control than does 1.5 kg/ha when broadcast as granules onto the paddy water. Furthermore, recent studies indicate that the whorl maggot can be well controlled if the rice seedlings are dipped into a mixture of gelatin, insecticide, and water—a method that requires only a small amount of insecticide and a little extra labor and equipment.

Additional research work is needed to develop a better applicator for placing insecticide in the root zone. The new chemical products being turned out by industry should be thoroughly evaluated to find compounds that are both effective and low in cost. Research on biological control should not slacken. There is evidence that the use of parasites—insects, bacteria, and fungi—as well as virus diseases that attack harmful insects may yet prove to be an inexpensive way of insect control.

Varietal resistance will continue to be the least costly way of controlling insect populations except for those species for which no resistant varieties have yet been found or when an insect species develops new biotypes so frequently that it is impossible for plant breeders to produce new varieties fast enough to maintain insect control.

The most likely approach to practical insect control in the years ahead is "integrated pest control," which consists of employing resistant varieties, cultural methods, biological control, crop rotation, or whatever method is appropriate for a given pest, and resorting to the use of insecticides only when those other methods have proved ineffective. It must be kept in mind that the use of insecticides is never profitable unless insect populations are high.

Better Weed Control Methods for Rainfed Rice

In the 1960s, scientists developed selective herbicides that controlled weeds in flooded lowland fields of transplanted rice. At that time, however, no herbicide satisfactorily controlled weeds under upland conditions or in any other circumstances where rice is direct-seeded without previous flooding.

In the 1970s decided progress was made in producing herbicides effective against many weed species in nonflooded fields. Nevertheless, there are a few weeds that are difficult to control, especially perennials such as nutsedge (*Cyperus rotundus*) and *Mimosa invisa.* The only satisfactory methods require two or three applications of several herbicides at various times, a laborious and costly procedure.

With the recent emphasis on growing two crops of rainfed rice in the monsoon season—the first one being direct-seeded before enough rain falls to permit puddling of the soil—it is becoming increasingly important to have herbicides that will kill all weeds and yet not be toxic to the rice plant. This is important, also, in flooded or deep-water rice fields, for they are usu-

ally direct-seeded after the first rains appear but before the flood
waters arrive. Frequently, yields are seriously reduced because
of competition from weeds during the early growth stages.

In flooded, transplanted rice fields, the surface layer of water
helps eliminate many weed species. Moreover, the transplanted
rice seedlings have a good start over the water-tolerant weed
species. In direct-seeded rice fields, however, the weeds and the
rice germinate at the same time, and there is no standing water
to inhibit weed growth. Therefore, the weed control problem is
difficult. Chemical herbicides likely will prove to be the most
economical way of controlling weeds in direct-seeded rice. To
weed a hectare of flooded, transplanted rice by hand may
require as much as 300 man-hours of labor whereas direct-
seeded, upland rice may require as much as 1200 man-hours. If
upland rice gets no weed control, the harvest is often reduced to
zero.

In summary, despite the important advances that have been
made in recent years, there is still great need for better selective
herbicides for use on direct-seeded, nonflooded rice that
depends on rain for its growth.

Fundamental Causes of Low Rice Yields

In 1976 the average rice yield for South and Southeast Asia—
specifically Bangladesh, Burma, India, Indonesia, Kampuchea, Laos, Malaysia, Nepal, Pakistan, Philippines, Sri Lanka,
Thailand, and Vietnam—was estimated by the FAO to be 1.81
t/ha. The 13 countries, with a total population of just over 1000
million, plant 86 million hectares of rice. Because the East
Asian countries (China, South Korea, and Japan) are not likely
to have much surplus rice in the years ahead, the South and
Southeast Asian nations will have to increase their rice
production steadily if they are to meet their own requirements.
That high yields are possible is shown by the fact that on
carefully supervised trials on farmers' fields conducted in the
Philippines in 1975, yields of 6.1 t/ha were obtained in the dry
season and of 4.6 t/ha in the rainy season. These yields are 1 to 2
t/ha higher than those obtained by the farmers themselves on
land adjacent to the trials.

It was fairly easy to identify the reasons for the differences in the carefully conducted trials. Because water control was good on the selected farms, 80 percent of the yield differences could be explained by the fact that more fertilizer and pesticides were applied to the supervised trials than the farmer normally applied himself. In fact, from the many studies that have now been carried out on farmers' fields throughout South and Southeast Asia, it has become clear that water control is the foremost yield constraint, followed by inadequate fertilizer use, and poor pest and weed control. Although the actual constraints vary from location to location, it appears that the more significant yield-limiting factors have been identified.

What seems necessary is a series of economic and social studies to find the basic reasons, for example, why adequate irrigation systems or flood control projects have not been constructed, why farmers are not using enough fertilizer, and why pest control or adequate weed control is not being practiced by enough farmers. The answers can lie in such factors as too low a support price for rice or an inadequate credit system. The studies generally would be conducted separately by country or by areas within countries. Such agencies as the World Bank and the Asian Development Bank are particularly well suited to conduct the needed surveys at national and regional levels.

Cropping Systems Involving Rice

Multiple cropping has been practiced in China for generations, and intercropping has been carried on by subsistence farmers in many countries for centuries, yet it is only recently that intensive research has been conducted in those areas of agriculture. The research projects are usually carried out under the general title of either cropping systems or farming systems.

Of particular concern to agricultural administrators in the less developed rice-producing countries of the tropics and subtropics is the possibility of annually growing several crops of rice or a crop of rice plus one or more other crops. Because Asia has little arable land left for expansion, in addition to raising yield per crop the only avenue left for producing more food is to increase the intensity of cropping throughout the year.

Three international agricultural research centers are conducting research on cropping systems involving rice. IRRI's cropping systems research program is devoted entirely to systems that include at least one crop of rice each year. Rice is of lesser concern at the International Institute of Tropical Agriculture in Nigeria or the Centro Internacional de Agricultura Tropical in Colombia. Therefore, only a portion of their cropping or farming systems research is concerned with rice as one of the crops.

IRRI recently inaugurated a cooperative program to conduct cropping intensity studies at 14 sites in six countries of South and Southeast Asia. Concurrently, it has initiated detailed studies at several locations in the Philippines. Essentially all of this work is being conducted on the farms of cooperating rice growers.

The principal objective of these studies in tropical Asia is to utilize more fully the water from the monsoon rains. Many farmers in monsoon Asia grow a crop of rainfed rice, wait until the soil is dry enough to cultivate, and then plant maize or some other upland crop. This causes an extended turn-around time. With this practice, any additional rainfall or residual soil moisture goes unused during the period following the rice crop. IRRI research workers have found that if two direct-seeded rice crops are grown, the first dry seeded and the second transplanted, and if those two crops are followed by a drought-tolerant crop such as maize or sorghum, much more food can be produced than under the present system. In several experiments on farmers' fields, as much as 10 t/ha of rice was harvested from the two crops of rice alone in years of normal rainfall distribution.

Additional work is needed on the techniques of direct-seeding the rice crop under both dry and wet soil conditions, on weed control methods under rainfed conditions, on ways of reducing the turnaround time between crops, on methods of insect control, on the use of fertilizer in multiple cropping, on the most effective way of handling crop residues, and on the profitability of supplemental irrigation in the multiple cropping system. As population pressures rise, multiple cropping will increase. This movement must be backed by sound and widespread research programs, with many trials on farmers' fields to aid in promoting rapid adoption.

Continuous Rice Production

Recently there has been considerable interest in designing systems of continuous rice production in irrigated rice fields of the tropics. This has been tried by scientists at CIAT in Colombia, by the applied research and training group at IRRI, and by an innovative Filipino farmer. All of these plans, although varying in detail, have been successful in providing the rice farmer a continuous source of income and an even distribution of labor throughout the year. In 1976, the IRRI initiated a research project based on a modification of a scheme worked out by a Filipino farmer who had produced 30 tons of paddy from a 1.5-hectare plot in one year. In the IRRI experiment, a 1-hectare field is managed by only three laborers. The field is divided into 250-square-meter plots. Every Monday, Wednesday, and Friday one plot is transplanted. On Tuesday, Thursday, and Saturday, rice is harvested from the plot planted 90 days earlier. Only 1 day of turn-around time is allowed, which means that the day after a 250-square-meter plot is harvested, the plot is again prepared for transplanting.

The system gives the farmer a steady year-round income, abolishes peaks and troughs in labor use, and maximizes annual yields per hectare. With an early maturing variety (IR36), IRRI obtained annual yields of 23.5 t/ha. To get such high yields management was intense: generous amounts of fertilizer and insecticides were used, and weed control was perfect. Nevertheless, economic analysis of the costs and income shows that the venture is a profitable one. The three laborers in the IRRI experiment, who work steadily for six 8-hour days, are former tenant farmers. They like the method, and some of their relatives are adopting it on their own farms.

This kind of experiment must be tried out more broadly and

over a number of years. Farmer reaction needs to be more fully explored, and the problem of disease and insect population buildup should be studied. Nevertheless, it shows great promise for use on irrigated land and by farmers who are willing to work diligently to increase their incomes.

7
Elements of a Successful Accelerated Rice Production Program

The production of rice during the past several decades has barely kept pace with population growth. In spite of the Green Revolution, then, the Asian rice consumer is no better fed today than he was 10 years ago. Yet, the yield levels of many rice-growing countries are only half of what they could be. To reach its full yield potential, each nation must mount an accelerated rice production program. This program must combine careful planning with sustained, energetic, and effective implementation. Especially necessary is the participation of government officials at all levels, from president or prime minister and minister of agriculture through provincial or state officials to the agricultural extension people in direct contact with the farmer.

Government involvement of this scope becomes more likely as the countries strive to improve their world financial position by limiting imports to nonindigenous commodities. Increases in the cost of fertilizer, farm machinery, and other agricultural inputs have been accompanied, understandably, by a rise in the price of food grains. Consequently, in less developed countries with meager foreign exchange reserves, interest in cereal self-sufficiency has mounted. For much of Asia, this means self-sufficiency in rice.

It is not essential, of course, for every nation to grow all the rice it consumes. Malaysia, for example, with its valuable exports of rubber, tin, and palm oil, can afford to purchase rice from outside. On the other hand, such countries as India, Bangladesh, Vietnam, and the Philippines, which are densely populated and have relatively low foreign exchange earnings, find it advantageous to produce enough rice to meet domestic demand and to divert foreign currency earnings to the purchase of the materials that are necessary for economic advance.

There are no simple answers to the problems of agricultural development. Requirements vary greatly from country to country, depending on the state of development of the nation. For instance, an accelerated production program for rice in Nigeria or Senegal, where up to now the crop has been a minor one, would require rather different guidelines, particularly in the early stages, than a program in India or Bangladesh, where rice has been grown extensively for centuries. Furthermore, the trade resources available for development differ considerably from nation to nation, some countries having oil or surplus agricultural produce for export and others having but few commodities with which to earn foreign exchange.

Despite such complexities, certain fundamentals of development apply generally to those countries in the tropics and subtropics where per capita incomes are low (often less than US$200 annually), and where a large segment (60 percent or more) of the population operates small farms, mostly of less than 2 hectares. It is to agricultural leaders in such developing countries that this and the following chapters are addressed. Although the information presented is not new, it nevertheless is a practical summary of what most authorities agree are the key elements in a strategy for mounting a successful accelerated rice production campaign.

But, first of all, it must be recognized that the outcome of a national accelerated rice production effort will not succeed unless administrative officials have a strong political will to achieve the desired results. Moreover, they must thoroughly understand it.

The distance between the seats of government and farmers' fields is normally a long one, both measurably and in terms of activities and objectives. Along the extended route from officialdom to the farming community, nevertheless, are those who, in varying degrees, are familiar both with national directives and with the more detailed workings of agriculture—those, in short, who can bridge the gap between the two. Only with a clear line thus connecting both ends of the endeavor will a nationwide crop production program succeed. In Taiwan and South Korea, for example, not only were top-level decisions made to mount an all-out effort to achieve self-sufficiency in rice production, but these were backed by a countrywide campaign, right down to the local level where every responsible farm family was expected to do its utmost to obtain higher yields than ever before.

Naturally, to be successful any operation functioning from the head of government to the farmer himself calls for a high degree of determination, dedication, and discipline at all levels. This can better be achieved, experience has shown, if suitable recognition is given to all who work effectively and coopera-tively for the program, and if all are kept aware of the importance of their part in it. The personal satisfaction to be gained from participating in a successful movement to increase food production (and thus contributing significantly, perhaps even historically, to the well-being of one's own people and country) is a morale-sustaining resource that should be recognized and skillfully used. It is particularly important that the farmers be given a chance to participate in decision making at the local level and be made to realize that, in the final analysis, they determine the success or failure of the national campaign to produce more food.

Assuming, then, that the energy and enthusiasm of a country's most promising personnel—specialists and non-specialists alike—can be marshaled to carry out a national drive for increased rice production, what factors must be taken into account in mapping out a successful strategy of operations? The first steps are to assess the country's natural resources as they apply to the growing of rice and to identify the important elements that constitute a successful rice production program.

Analyzing the Natural Resources

Water Supplies

No factor is more critical to rice production than water control. In planning a national program of accelerated rice production, it is essential to have a thorough knowledge of the country's water resources, both from streams and from ground sources. If nationwide surveys have not been conducted, they should be arranged without delay. If the capability for making such surveys is not available within the country, international aid agencies (see appendix) can assemble teams for that purpose. In addition, of course, there are numerous commercial firms in North America, Europe, and Japan that can be hired to make surveys of water resources.

Administrators and decision makers must find out the extent to which nonirrigated lands, or areas that are now only partially irrigated, can be provided with a year-round water supply. Furthermore, of course, they must have information on how such potential sources can best be developed. Water resource surveys tell the planners whether irrigation water can be obtained most dependably and economically in a locality by building reservoirs, by placing barrages in the rivers, or by tapping groundwater sources. If the latter, they need to know whether the wells should be deep or shallow, and what the long-range predictions are for a continuing and adequate supply of water.

Although a lack of water is the primary constraint to high and stable rice yields, too much water is likewise a problem. There are, for instance, vast areas in South and Southeast Asia where floods recurrently damage rice during the monsoon season. Here again, the advice of experts is needed to determine the feasibility of drainage or flood control projects.

Irrigation projects are expensive, but when properly engineered and managed, they give good returns on the investment. The Asian Development Bank states that, in six of the large irrigation projects that they have supported in Asia, the average investment cost was US\$1500 per hectare. Justifying that outlay was the subsequent increase in annual rice production on those projects of 5 t/ha, resulting mainly from the fact that two crops a year could be grown where only one was grown before.

Studies in Southeast Asia show that the success of irrigation projects lies in their proper design and management. The capacity of the system must be adequate to supply all farmers in the command area. Too frequently, in their anxiety to serve as many farmers as possible, irrigation agencies spread the water resources over an excessively large area, which greatly intensifies the competition among farmers for an adequate supply. In such circumstances, the farmers at the head of the main canal tend to use more than their share of water, leaving little or none—especially during the dry season—for those farther along the system.

Surveys among Filipino and Indian farmers reveal that if sufficient water is available and if the system is carefully explained to the farmers, they have good cooperation among themselves and are willing to pay the irrigation fees. To gain the farmers' confidence, however, the system must be managed by competent field men. There is little point in trying to supervise farm-level distribution in lateral canals properly unless the main canal is equally well regulated. The formation of farmer-controlled irrigation associations, which work closely with the irrigation authorities at the national or district level, is often the key to successful water management.

Where adequate groundwater resources exist, many farmers prefer to sink tube wells, chiefly because they are then able to control their own water supplies. Generally, however, tube wells are expensive sources of water. There is not only the initial cost of drilling the well but also the continuing expense of diesel fuel (or of electric power where available) to pump the water. Still, thousands of deep wells have been drilled in the Indian subcontinent during the past 20 years, most of them privately funded and thus not requiring government support.

Soil Conditions

Although water control is a major factor determining the feasibility of increasing rice production, some attention must be given to soil and topographic conditions. Sandy soils are usually unsuitable for rice growing. Their low capacity for holding water and nutrients and their high permeability make it difficult to maintain the necessary flooded conditions without using excessive quantities of water. Soils of this type normally can be put to better use than for rice production. Sweet potatoes, for example, can often be grown satisfactorily on sandy soils where rice would give deplorably low yields.

Of the acid-sulfate clay soils, some can be so managed as to produce reasonably good rice yields, but many of them are so acid and have such a high acid reserve that they cannot be reclaimed economically. Clay soils that are not too acid, of course, are ideal for rice.

Hilly lands are not well adapted for rice growing. Unless they are terraced and bunded so that water can be stored during periods of abundant rainfall, yields will be low and extremely unstable. Even with terracing, a minimum rainfall of 1500 millimeters, largely occurring during the growing season, is needed for reasonably dependable harvests. Moreover, terracing is expensive and is seldom advisable at present-day costs of labor and equipment. Although there are extensive areas of upland rice planted on hilly land, without exception those countries growing a high proportion of rice under upland conditions have national average yields considerably below 2 t/ha.

Low-lying flat lands with clayey soils, on the other hand, are ideal for rice production. They not only are most suitable for the special requirements of the rice plant but, particularly during the rainy season, are largely unfit for any other major food crop.

Climatic Conditions

Rice performs best between about 40 degrees north and 40 degrees south of the equator. Growing rice at latitudes more than 40 degrees requires special early maturing varieties and the use of plastic-covered seedbeds. Always, too, there is a hazard of early frosts at flowering time and consequent severe crop losses. Similar problems confront the farmer attempting to grow rice in the tropics at elevations above 2000 meters. For those reasons only a small fraction of the world's rice is grown at high latitudes and altitudes.

Pest Populations

Certainly not a natural resource (although possibly an "anti-resource"), the pest populations of a country are a natural presence influencing the success of a rice production program, and consequently they should be considered in the planning process. An analysis of previous pest outbreaks and a survey of current pest control problems are important in determining what methods of attack should be planned. Assessment should be made of such factors as the amount of rat damage that occurs, the depredations of migrating or indigenous bird populations, and of course the kinds of insects and the extent of their damage.

The Essential Elements

A Problem-oriented Rice Research Program

The biological components of agricultural technology are not wholly transferable from one region to another. Environmental conditions that vary widely dictate the usefulness of certain materials, methods, and practices, and the impracticability of others. This means that every rice-producing country needs its own program of rice research, which includes the testing of genetic lines and rice varieties developed by the international agricultural research centers or by the large national rice improvement programs that carry on comprehensive breeding research. In addition to such plant testing, the regional research stations will find it useful to conduct soil fertility trials to determine optimum fertilizer treatments, to work out appropriate insect and rodent control programs, and to study cultural practices suitable for the soil and general environmental conditions in the area.

Major rice-producing countries can fully justify a complete research program incorporating plant breeding, entomology, plant pathology, plant physiology, agronomy (including soil science), agricultural engineering, and agricultural economics. India is a good example of such a country. It has not only central rice research stations operated at the national level but also many state-controlled research programs associated primarily with agricultural universities.

It is safe to say that any nation growing more than 200,000 hectares of rice is justified in supporting one or more rice experiment stations that conduct adaptive research. This is in addition to on-farm trials and demonstrations. Obviously, the research stations conducting problem-oriented rice investigations should be located in the principal rice-growing areas, where the results will be more applicable to the surrounding farm conditions.

On-farm Trials and Demonstrations

Even though a country has a well-run, problem-oriented rice research program, it is still of utmost importance that on-farm trials of varieties and of cultural practices be conducted. These trials serve two important purposes. First and foremost, they provide a means of determining the suitability in each locality of the more promising varieties and the accompanying modern practices. Second, they permit local farmers to observe the results. Because the preliminary screening of varieties and testing of practices will have taken place at the various experiment stations, most of the trials will be successful. Consequently, they will double as convincing demonstrations. This observable example gives the farmer confidence in the new technology—perhaps to a greater extent than does any other method of persuasion.

On-farm trials should be conducted jointly by scientists from research stations and by the extension staff to facilitate contact and the exchange of ideas between research and extension workers. In addition, the research scientist benefits from seeing his findings tested under farm conditions, where results may be quite different from those he obtained at the experiment station.

On-farm trials should be replicated, frequent observations should be made during the growing season, and, of course, yield data should be obtained. Although they may prove to be valuable demonstrations, their primary purpose, it must be kept in mind, is to test promising varieties and methods under farm conditions.

The number of trials to be established within a locality or district depends on the size of the area being covered and the variability that exists within it. If it has significantly different soil types, fertilizer trials should be run on each major type. If sizable variations in elevation exist, separate varietal trials should be conducted for altitudes that differ by more than 250 meters.

As results are obtained from on-farm trials and as research

and extension personnel gain confidence in certain varieties and methods, demonstration plots should be set up on farmers' fields. Demonstrations have a different objective than on-farm trials. They would not be replicated on any one farmer's field but would be set up on the fields of many farmers. No large rice-growing village should be without one or two demonstrations.

The demonstration plots should be simple. The idea to be introduced depends upon the level of technology in the locality. In some areas, a simple weed control demonstration together with an improved variety may be appropriate; in others, where modern varieties already have been introduced, treatments to show the advantage of fertilizers or insecticides may be called for. The demonstrations should vividly portray the response that can be obtained with the use of the technique or input being featured. As extraneous factors in the test plots can ruin the demonstration, the extension agent in the locality must closely supervise the plots, not only when they are first established, but during the entire period that the rice crop is growing. Poor weed control is an extraneous factor that can make a fertilizer demonstration ineffective; birds and rodents, which can destroy any test, must likewise be guarded against.

The value of demonstrations on farmers' fields cannot be overemphasized. A farmer who sees a successful test on his land or his neighbor's will be far more willing to change from the traditional methods to the modern ones. Indeed, there are no known examples of rapid progress having been made in the introduction of modern rice-growing practices in the absence of on-farm trials and demonstrations.

Availability of Inputs

If modern varieties are to be grown successfully, needed inputs must be available to farmers. The importance of water control has already been stressed. Weed control is essential, but most farmers realize this and do a reasonably good job of hand weeding. Therefore, under average conditions, if water supplies are adequate, the next most important input is fertilizer.

Fertilizer. Although there are some fairly young (geologically speaking) volcanic ash and alluvial soils in the

Philippines, Indonesia, and elsewhere on which rather high rice yields can be obtained without fertilizer, yields of over 4 t/ha ordinarily cannot be attained without fertilizer, and unfertilized fields usually yield much less than that. Even the more fertile soils, if intensely cropped, will soon need fertilizer, because each rice crop removes substantial amounts of plant nutrients.

If applied generously, organic fertilizers (animal manures and compost) are able to produce good rice yields. However, since vast areas of the world's rice land cannot be supplied economically with sufficient quantities of organic materials to support high yields, the only alternative is to use chemical fertilizers. (Even in China, where the traditional use of natural fertilizers, including night soil, continues to be extensive, chemical fertilizer factories are operating at maximum capacity, and building new factories has become a priority goal.)

Fertilizer application should never be haphazard. Before planting time, stocks of chemical fertilizers must be available at the places where they are to be used, in the amounts needed, and at a fair price in relation to that of rice. The kind and amount of fertilizer most suitable for each area can be determined by adaptive research and on-farm trials. With the exception of a few special environments, rice will always respond to nitrogen (provided water supply is adequate and insects and weeds are controlled). In many regions, phosphorus, in addition to nitrogen, is required for high yields. On certain lighter textured soils or on older soils, potassium may be limiting. In each area a test should be run to determine whether zinc may be needed.

Production and distribution of good seed. The significant advances made in breeding superior rice varieties during the past two decades make it highly important that pure seed be made readily available to farmers. Too frequently, there is no adequate system for multiplying the seed of new varieties and for distributing it to farmers. In some countries it takes 10 years from the time a cross is made until one of the progeny is released as a variety and seed is available generally in the countryside. Elsewhere this same process has been accomplished in only 4 years. The rapid change in varieties that occurred on farms in Taiwan, South Korea, Colombia, and the Philippines, for example, would not have been possible without efficient seed programs.

Too many countries are unnecessarily cautious with on-farm trials and unduly restrict the release of promising genetic lines and varieties. In addition, there are often not enough seed farms, either private or governmental. As an alternative, progressive farmers can be taught to produce good seed. Rice, being a self-pollinated crop, is relatively easy to manage for pure seed production.

Every country should have a seed certification program, so that top-quality seed is available to farmers who wish to purchase it. Furthermore, it is important that a source of pure seed of the leading varieties be maintained. A good seed certification program requires thorough organization and control. It starts with breeder's seed, followed by registered seed, and finally by the certified seed that is actually produced on private and government seed farms, under rigid inspection in the field and at harvest time.

In spite of the importance of a national certified seed program, certified seed need not be planted by every farmer who wishes to grow the modern varieties. Even in developed countries, many rice farmers grow what is termed "good seed," which would not quite pass the rigid standards for certified seed.

Usually the local agricultural extension officer is the key person in determining whether there is a problem in getting good seed to the farmer. If there is, the difficulties should be identified and corrected. With proper communication among plant breeders, research administrators, and the government agencies responsible for seed multiplication, certification, and distribution, good seed of the appropriate modern varieties should be available to all rice farmers in any country.

Insecticides. Although insecticides are sometimes necessary, they should be used only when other methods have proved inadequate. The kind and quantities of insecticides farmers need vary from country to country, depending on the prevalent insects. Detailed information in regional rice production manuals and various extension leaflets serve as the best guides to agricultural officials in the various countries and localities. Furthermore, in most places the commercial companies that sell or manufacture insecticides provide ample information on the use and the effectiveness of their products.

Herbicides. There have been tremendous advances in the production of new herbicides that effectively control weeds in rice fields. Undoubtedly, the use of chemical herbicides will increase as time goes on. To the extent that there is a demand for them, they should be provided.

The use of herbicides is not essential, of course, for weeds can be controlled by hand methods. Whether or not to use chemical methods of weed control is a matter of timing and of economics. Studies in Southeast Asia show that often farmers are so late in removing weeds from their fields that the crop is already damaged by the time the last of the weeding is done. When labor is either scarce or costly, herbicides may prove to be profitable. Either way, good weed control is imperative if fertilizers are to be used profitably.

Power equipment. Farm power equipment, especially power tillers, four-wheeled tractors for land preparation, and portable grain threshers, will be in greater demand as more rice land is irrigated and the consequent opportunities for double and triple cropping increase. The use of gasoline- or diesel-powered equipment in land preparation becomes profitable when two or more crops are grown annually on the same land. The turn-around time is much shorter with power equipment than with animal power.

Direct-seeding equipment, small grain dryers, seed cleaners, and simple low-lift water pumps are available. In the more industrialized Asian nations, mechanical transplanters are becoming popular among farmers. An assessment of the degree of development of a given rice-growing area should indicate to agricultural officers the extent to which mechanical equipment would be in demand. Naturally, in the early stages of development other inputs will have a much higher priority.

Agricultural Extension

Many problems remain for research scientists to resolve. Nevertheless, the bottleneck in raising rice yields on farms is often a weak extension program. Too frequently, extension workers cannot advise farmers properly because they are inadequately trained. Often they are given assignments not directly related to increasing crop production. In some countries, although large numbers of extension workers are assigned to the field, their operational budget is meager, and they have no means of transportation. As a result, extension workers are not able to visit the farmers on a regular basis; instead, farmers have to seek out the extension officer at his headquarters.

Extension programs have been organized successfully in various ways in different countries. Sometimes extension is carried out through farmers' associations or farmer cooperatives. Or it may be part of the program of the national community development agency. In numerous countries it is a function of the ministry of agriculture. Many authorities prefer this arrangement, for they feel that increasing crop production is the main purpose of the agricultural ministry and that the varied functions of some of the other agencies prevent them from giving an accelerated crop production program concentrated attention. A few nations, such as the Philippines and Indonesia, have developed special accelerated rice production programs into which the major rice-extension activities have been incorporated.

Regardless of which agency operates the extension service, certain principles apply to any successful extension program. The principles described briefly below are based on Benor and Harrison's *Agricultural Extension: The Training and Visit System,* which is a helpful guide to a sound and well-organized extension system.

Training the extension staff. Although no component of a rice extension program is more important, surely, than a well-qualified field staff, too often extension workers are inadequately trained. Although it is valuable and useful to learn the principles of communication, by itself even the most thorough knowledge of that discipline is insufficient preparation for extension personnel. People who are in constant touch with the rice farmer must fully understand how to grow a good crop and how to diagnose the problems a farmer may encounter. Unless an extension agent has had an opportunity to grow a rice crop himself, he is unlikely to be of much help to the farmer. On the other hand, if he has spent 6 months in a rice production training course that provided him an opportunity to perform all the field operations necessary in growing rice, from planting to harvest, he can face the farmer with confidence and guide him effectively. (Rice production courses, described further in the appendix, are available in Asia, Latin America, and Africa.)

Training of extension people can be a continuous process. In the "training and visit" system, a training session for field workers takes place every 2 weeks, mainly to instruct them in certain timely techniques that will be brought to the attention of rice farmers during the following 2-week period. If possible, however, all field workers should go through a basic rice production course before they take up their duties as rice advisers to farmers. For those who cannot be spared for 6 months, there are 2-week courses available and even in that short time a trainee can learn much about modern rice production methods.

Within the limitations of basic training, naturally, a field technician cannot become an expert in the many problems affecting the rice plant. Essential to the success of every rice extension program is the availability of specialists—in such fields as plant pathology, entomology, and soil science—to identify diseases and insects and suggest appropriate methods of control, to provide the remedy for any unusual adverse soil condition that may exist, and to advise in other specialized areas.

These specialists could come from the faculty of an agricultural college, from the ministry of agriculture, or from some international agency, depending upon how agricultural research and extension are organized in a country. The important thing is that they be readily available to the extension field-worker when he encounters a baffling problem.

It is desirable for these experts to help the extension field staff, thus affording an opportunity for mutual acquaintanceship from the start.

Scheduled visits to farmers in their villages. No matter how well trained an extension worker is, he will be of little value in conducting an accelerated rice production program unless he visits farmers frequently. To do so, he needs transportation. The kind of vehicle would depend on the locality: in some, a bicycle would suffice, in others a motorcycle would be needed.

In some successful extension programs, a firm schedule is followed. Usually each extension employee is detailed to visit every rice-growing village in his area once every 2 weeks. The farmers then know that he will be in their village on the same day of the week every fortnight. Although the extension worker cannot call on every farmer during those visits, he is available to any of them who have serious problems and who request help. It is advantageous for the extension worker to select two or three key farmers in each village who he expects will adopt new techniques rapidly and who will pass information on to other farmers in the village. He should call on the selected farmers at each visit to acquaint them with any new or timely information he has. The farmers quite naturally may have a worthwhile experience to report to the extension worker as well.

Field days. Assuming that thriving on-farm trials and demonstrations exist in the area, it is useful to schedule farmers' field days. The focal point is usually either the tests and demonstrations or the fields of one of the selected key farmers who has an outstanding crop to display. Generally, a number of villages are served by a single field day. It is under such group conditions that many farmers are persuaded to change varieties and other practices. Nothing is more convincing to farmers than witnessing success in fields and surroundings they know intimately.

Unified extension service. Because conditions vary so much from country to country, it is not feasible to attempt to specify the actual organization of an extension service, from the top officials in the central government to the extension workers in the villages. Nevertheless, the more successful extension programs generally have had a single line of command from the governmental agency responsible for agriculture right to the field staff. Unless the head officials are convinced that an active and well-supported extension service is essential for progress, support for the program at the farm level will never be adequate.

Sidē roles in extension. Ideally, extension field workers should be occupied purely with promoting an accelerated rice production program among farmers (or with a multiple cropping program in which rice is the main crop). The extension worker, however, must be concerned about all the ingredients necessary for progress in production (fertilizer supplies, credit sources, and the like), and he must use his influence to see that any missing ones are provided. He may have to approach the agencies responsible and point out that the program is being held up by the unavailability of inputs and services. On the other hand, it would be unfortunate if, for example, he had to run around the countryside trying to locate fertilizer supplies for individual farmers.

On occasion, too, the extension agency cannot avoid giving the field staff certain responsibilities that are not strictly associated with advising farmers in crop production techniques. For instance, in the Masagana 99 program in the Philippines, the extension workers help farmers fill out loan applications and do a certain amount of loan collecting as well. Philippine officials justify this on the basis that credit is essential for the success of the program. They add that because the farmers are often illiterate, the field staff, all graduates of agricultural colleges, are able to help in preparing loan applications that are more likely to be accepted by the lending agency. Moreover, in the process of helping farmers apply for loans, extension workers frequently gain their friendship and confidence and, as a result, can often be extremely successful in getting the farmers to repay their loans at harvest time. Another argument for involving the extension staff in rural credit is that it provides additional occasions to visit the farmers and their fields and to discuss any problems they may be having.

Maintaining staff morale. Generally speaking, extension service employees have a lower status than research workers. Often they are paid less, and low staff morale can be a serious problem. One way to improve morale is to recognize the top achievers. When a field staff member is notably successful in getting agriculture moving in his locality, he should be praised publicly. This is the responsibility of district and national officers when they visit the extension projects. Through such genuine efforts to let the field staff know they play a critical role in agricultural development, the extension worker gains confidence in his ability, develops pride in his work, and makes an extra effort to attain further success. In addition, of course, everything possible should be done to pay extension workers well and to provide merit increases for those who turn in exceptional performances.

The adequacy of funds and materials strongly affects staff morale and achievement. If sufficient funds are not available for travel, for visual aids, and for the materials needed to put on vivid demonstrations of new techniques, it is usually better to reduce the size of the staff and properly support those who remain.

Farm-to-Market Roads

For the foreseeable future, the market for rice in Asia and Africa will be good. Unless farmers have ready access to those markets, however, they will have little incentive to raise rice production beyond subsistence levels. Furthermore, inputs (particularly fertilizer) can reach farmers only if there are roads into the villages capable of accommodating four-wheeled vehicles. Rural areas of South and Southeast Asia are dotted with isolated villages where development has been thwarted by the lack of roads capable of accommodating trucks. In many areas, dirt roads are passable only in the dry season; during the monsoon rains, even ox carts may become mired.

This backwardness can be changed. With proper inspiration and leadership, a village can build its own access road on a community self-help basis. In the off-season, farmers using their own oxen or water buffaloes, and with volunteer labor from other people in the village, can construct an all-weather road for the few kilometers from the town center to the nearest main highway.

Rural Credit

If progressive agriculture is to be developed and maintained, production credit must be extended to farmers. The formal farm credit systems include private commercial banks (including rural banks), farmers' associations, cooperatives, and various types of government lending agencies. These agencies are usually backed by the government or by the central banking system of each country. They are well equipped to supply credit to the more progressive, better educated farmers who are commercially oriented; in fact, they have been doing so for many decades.

The principal problem faced by lending organizations is how to accommodate small farmers who have limited resources and little collateral. Such farmers tend to fall into two classes: (1) those whose resources are so low that they cannot support their families even at subsistence levels, and (2) those whose resources, if properly used, are sufficient to allow them to make a respectable living but who are poverty stricken because they lack the credit to switch from traditional to modern agricultural methods.

The farmers in the first category are not good credit risks unless their family incomes are augmented by off-farm employment. Those in the second category, however, should receive major attention from the formal lending agencies, for only with such financing will they be encouraged to adopt the new rice technology with its attendant input costs.

In many countries, the number of production loans to small farmers has increased markedly in recent years. Unfortunately, however, the rate of repayment of such debts has been disappointing. Some rural banks and other lending facilities have simply refused to make a second loan to any farmer who did not repay the first. This has caused such farmers to revert to borrowing from the private moneylender, who, though more flexible in his lending policies, usually charges exorbitant rates of interest.

The *Asian Agricultural Survey 1976*, published by the Asian

Development Bank, lists three desirable characteristics of a small-farmer credit system. First, loans should be made available to farmers in time to meet the expenses of crop production. Second, repayment should be deferred when a farmer has suffered from crop failure or other unforeseen calamity. Third, credit agencies should be prepared to deal with large numbers of very small farmers in both lending and repayment operations.

Small farmers frequently complain that loans are finally approved only after the planting season is over—when it is too late to use the funds to buy fertilizer and other inputs that would increase the farmers' rice yield and provide the extra income from which they could repay the loan at harvest time. Such late arrival of credit further tends to cause farmers to use the money for the purchase of consumer goods that are not related to increasing the income from their land.

Too few lending agencies have sufficient staff to supervise credit properly among their clients. In the Philippines, those rural banks that added extra staff to supervise loans had a much higher collection rate than those that tried to do it with an inadequate number of field men.

In addition to such obvious factors of indebtedness as crop failure or the misuse of credit for nonproductive goods, too many small farmers neglect to repay their loans simply because they feel that governments and banks are "rich" and that therefore they do not need to return the money they borrowed. This problem is not easily solved, but the answer seems to lie in educating the farmer to understand (1) that no lending agency can continue to operate unless it collects its loans and (2) that without such agencies he will have no way to obtain credit except at usurious rates. Part of the difficulty can be solved by making loans in kind rather than in cash. There is then greater assurance that the loans will be put to productive use. In the end, lending and repayment policies must be tough, yet sufficiently flexible to provide noncollateral production loans to deserving small farmers.

Most government-sponsored credit programs charge low interest rates in the belief that this encourages the small farmer to borrow the needed funds. Studies of this policy indicate, however, that as long as interest rates are reasonable, the small farmer will continue to borrow. Furthermore, low interest rates encourage the larger farmer with a good business sense to borrow at "bargain" rates, to the partial exclusion of the needier farmer.

Another weakness of the low interest rate is that small-farmer loans become unprofitable for rural banks and other lending agencies, which in turn makes them reluctant to hire enough loan supervisors to achieve a good collection rate. It would seem more logical to maintain interest rates at a sufficiently high level to permit the lenders to service their loans efficiently and improve collections. The interest rates charged would still be significantly lower than those of private lenders.

Although the credit needs of most small rice farmers continue to be met from such informal sources as the private moneylender and landlords, it is extremely important that government-backed lending organizations develop viable credit systems for these growers.

Price Incentives

Two principal price policies favorably influence production and farm incomes: controlling the price of rice and subsidizing the cost of inputs. Obviously, governments can use a combination of these two policies, and some do. Because economists do not agree that one method or the other is superior, and because there are marked differences in the economies of the less developed nations, a firm recommendation cannot be made as to which is preferable, but some of the advantages and disadvantages can be listed.

Controls on the price of rice. The price of rice is controlled in two ways. One is simply to set it well above the world price. The other method is to set a minimum guaranteed price level. If the price on the free market drops below that level, the government agrees to buy the rice.

The first method (used in Japan) provides a strong inducement to production but is so costly that most of the less affluent, nonindustrialized nations cannot afford it. The costs to the urban dweller are high, and the government usually must subsidize the cost of rice to the consumer, too. In the poor countries, this policy consumes government funds that are sorely needed for investment in the infrastructure, for increasing wages, and for achieving other goals. Furthermore, high rice prices favor the large farmer who has an abundance of rice to put on the market more than the small farmer who is growing rice chiefly to satisfy the needs of his family.

In most tropical rice-growing countries where 60 to 75 percent of the population is engaged in farming, it appears better to set a minimum price for rice at a reasonable level. This gives the farmer confidence that regardless of the abundance of the rice harvest, he will get a good price. The government is protected because the price is not exorbitantly high, and in poor crop years the market price will be above the support price. In recent years, that policy has been followed in Taiwan and seems to be working well.

Although the guaranteed price for rice may have to be set annually, two important principles seem clear. One is that once a government decides to support the price of rice at a minimum level, it must continue to do so year in and year out. This makes the farmer feel secure in his investment of time, labor, and valuable inputs, and he will not be seeking alternatives to growing rice. The other principle is that the guaranteed price should be high enough to allow the farmer to purchase the necessary inputs and still make a profit. In other words, as the cost of inputs—such as fertilizer—increases, the price of rice should rise accordingly.

Subsidizing inputs. Many governments of rice-growing countries subsidize inputs. In few countries do farmers pay the full cost of irrigation water, for example. Fertilizer is also a commonly subsidized input. Less frequently, farm equipment, electricity, and insecticides are provided to farmers at less than cost.

In times of abnormally high prices (as, for example, in 1973-74 when the price of fertilizer soared), there is justification for subsidizing inputs; but in normal times, and if there is an appropriate minimum support price for rice, it is difficult to argue in favor of such subsidization. Nevertheless, many authorities prefer subsidizing the cost of agricultural inputs to providing support prices for rice. Proponents of subsidized inputs claim that the lowering of the price of an input to the farmer makes it more certain that he will use that input, whereas higher prices for the crop do not give any assurance that the increased income will be used to purchase materials that will enhance yield. It is contended, also, that subsidizing inputs, unlike increasing the price of rice, does not affect the incomes of the urban and rural poor who produce no rice. A third argument for input subsidies is that they can be rather flexible and thus can be varied in accordance with the economic levels of specific farming regions.

Actually, general input subsidies have had little impact on production. Perhaps this is partly because the use of expensive inputs in the less developed countries is still a small part of the total cost of production, especially among small farmers. There are numerous examples of the misuse of subsidized fertilizer; sometimes the poorer farmer sells it to the more affluent farmer. The flexibility argument is nullified further by the consideration that when fertilizer, for example, is subsidized for use on rice and not on, say, sugarcane, the rice farmer might sell it to the cane grower at a higher price than the subsidized one. Despite these difficulties, the subsidization of agricultural inputs may be justified in certain countries or under special circumstances.

Off-farm Employment

The hungry people of the world are the poor. It is now recognized that regardless of advances in agricultural science, the world food problem cannot be solved unless the problem of poverty is attacked concurrently. Although the poor and undernourished in the less developed countries are found in both rural and urban areas, because such a high percentage of the population is engaged in agriculture, the majority of the low-income people live in a rural environment. The causes of rural poverty are many, but overpopulation, the small size of farms, low crop yields, and a general lack of off-farm employment opportunities are the chief factors responsible for the excessively low incomes in rural areas. Moreover, the rural population in most less developed countries has been able to contribute little to the national economy, mainly because the majority of farmers are operating at subsistence level and have little cash to spend. Since the greater number of people in low-income countries are engaged in agriculture, economic development cannot proceed unless the earning and purchasing power of that huge segment of the population is increased.

It is often suggested that rural poverty can be avoided by using labor-intensive farming methods instead of mechanization. This course has some validity as long as the productivity of the land is increased in proportion to the expanded use of labor. However, there are limits to yield, and hence there is a limit to the number of people who can be fed and supported from a given area of land. The policy of continually increasing the farm work force is bound to exacerbate rural poverty and to result in economic stagnation. The only long-term remedy for the situation is to provide nonfarm employment opportunities in agricultural areas. The creation of alternative employment in farming communities particularly helps landless laborers and the rural unemployed, whose numbers are ever on the rise. Furthermore, it reduces the tendency for rural people to seek employment in the already overcrowded cities.

There are two principal strategies for creating off-farm jobs on a large scale. One is the establishment of labor-intensive public works projects. The other is the development of manufacturing industries in the rural areas. Programs to create more jobs are especially needed in the rice-growing regions, which are densely populated and contain many unemployed and underemployed people.

Labor-intensive rural public works. The kinds of rural works projects that are appropriate depend on the needs of the region. Common deficiencies are farm-to-market roads, bridges, and irrigation and drainage systems. Generally, projects that directly benefit the communities in the area are the easiest to accomplish, because they attract local support and cooperation. Usually, however, in addition to community support it is necessary to obtain outside funding, which may involve foreign aid or at least support from the national or provincial government. Any rural public works program should be a permanent scheme, moving from one project to another through the years, rather than a relief measure soon to be abandoned. The problems of rural poverty are not overcome by short-term programs.

Rural manufacturing industries. As a rule, the most suitable rural manufacturing plants are small and labor intensive. Although considered to be non-urban, they should be near secondary cities or market towns. If indiscriminately dispersed through the countryside, electric power, communication and transport facilities, banks, and the like may be lacking.

For areas that have yet to develop rural industries, the most appropriate are agro-industries, including factories both for food processing and for manufacturing agricultural machinery and equipment such as power tillers, irrigation pump sets, animal-drawn equipment, and hand tools. Modern fertilizer factories are not small-scale industries, but the bagging and distribution of fertilizer are operations that can employ many rural people. Countries with abundant coal resources might find it advantageous to manufacture ammonium bicarbonate in modest-sized plants of the type widely found in China.

As the need for the products of agro-industry is met and further expansion is thus unwarranted, the region can proceed to establish plants for manufacturing consumer goods such as clothing, furniture and other wood products, and plastics.

Many less developed countries have made a start in providing off-farm employment in rural areas, but for the most part such efforts must be intensified if the job opportunities are to keep up with population growth.

8
A National Rice Program:
Putting the Ingredients Together

In presenting the more important elements of a successful rice improvement program in the preceding chapter, no attempt was made to rate them according to their importance or to the difficulty of introducing them into a rice production scheme. To do so would be arbitrary and theoretical in view of the interdependence of the program ingredients and of the differences among countries in stage of development and in natural, social, and political environments.

The interdependence and interaction among the various ingredients must be appreciated. For example, if irrigation water is available but no fertilizer is used, yields will remain low. If modern rice varieties are not planted, there will be little response to the use of fertilizer. If a workable rural credit system is lacking, farmers cannot raise the money to purchase fertilizer and other supplies needed to increase yields. If production incentives, such as a minimum guaranteed price for rice, are not available on a permanent basis, farmers may be reluctant to invest in the inputs required to express the full yield potential of the modern varieties. Thus the list continues, each element of a successful rice development program inevitably linked to others.

Nevertheless, decisions do have to be made as to which components should receive primary attention. Administrators and planners endeavoring to increase a nation's rice output should first examine each geographic area being considered for improvement to determine its state of development with respect to each of the elements needed to implement an accelerated rice production program. It is essential to identify constraints to yield, whether technological, economic, or social. The decision then to be made is which obstacles are the most limiting and to what extent available resources will permit them to be removed or at least to be markedly reduced.

This final chapter offers guidance in assessing the rice production potential of an area, in selecting and undertaking steps to achieve that potential, and in maintaining the increased pace of rice production after it has been set. Described first is a rural structure of the type needed to facilitate the flow of goods and services in the countryside—a structure that includes the major elements set forth in chapter 7.

Organizing the Rural Structure

Once the decision has been made to mount an accelerated rice production program, attention should be given to the organization of public and private services at the local and district levels and to the relationship between those sectors and operations at the national level.

The Farming Locality

Subsistence farmers can operate reasonably well without much dependence on outside organizations or agencies, governmental or private. However, when a farmer moves from the subsistence level to producing a marketable surplus, he immediately requires various services. He needs technical advice on improving crop yields, and he must have access to input and output markets, and to credit facilities. Such services should be available within a reasonable distance of each farmer. To achieve this proximity, an area has to be divided into rather small units, each of which contains the assemblage of facilities.

A. T. Mosher terms these units "farming localities," a usage that is followed here. A farming locality is similar to the "community" as defined by early American rural sociologists. In China, the "commune" provides the facilities and services that would be contained in a farming locality. In Taiwan, the farmers' associations substitute for the farming locality.

The size of a farming locality depends upon the quality of its transportation. The farming locality should be small enough so that the average farmer can travel readily from his home to the market center and back in a day. Farming localities tend to become larger as development takes place. If, for example, most rice farmers have no transport beyond a buffalo-drawn cart, the radius of a farming locality should not be greater than 5 to 7 kilometers. As access to trucks and buses becomes available, the locality can be considerably larger.

As a minimum, each farming locality should contain: (1) a market center for selling rice and for purchasing farm supplies; (2) rural access roads to connect the market center with the outside world, in addition to roads leading to the market center from the villages within the farming localities (the quality of the roads should be in keeping with the kinds and the amount of traffic they have to accomodate); (3) adaptive research trials and on-farm demonstrations; (4) an extension agent who is qualified to help farmers manage modern rice varieties; and (5) some sort of credit office (usually part of a large national organization) in each market center.

The Farming District

The "farming district" is a larger unit. It serves the farming locality as the latter serves the farmer. The facilities and personnel in individual farming localities are not self-sufficient. They must be tied to larger units in a nearby city. Mosher describes the farming district as the truly basic unit for creating a progressive rural structure. It is the smallest unit that can afford to have all the services for agricultural development in an area.

The size of a farming district depends on such factors as the density of the farming population, the topography of the area (long narrow valleys, for instance, that grow lowland rice almost exclusively would have a different distribution of farming localities and districts than would a broad expanse of flat land not limited by hills), and the presence of secondary urban centers. Frequently a farming district would include from 20 to 30 farming localities.

The heart of a district is obviously an already existing

population center of larger size than any community within
the group of farming localities that it serves. In areas where rice
is the major agricultural commodity produced and sold, the
district center probably would be located on flat land
surrounded by extensive areas of lowland rice. The center of the
farming district normally provides wholesale markets, a
regional rice research facility, a central extension office, district
banks, and communication links to the farming localities
within the district.

Wholesale markets are needed in the district centers, because
rice seldom can be moved directly from the market centers in
the localities to the nation's main urban areas where much of it
finally is sold. As part of the marketing system, the district
center should also contain sizable rice mills and storage
facilities. Furthermore, relatively large distributors of farm
equipment and supplies would be established in the district
center. Such companies would be able to supply fertilizer,
machinery parts, insecticides, etc., to the smaller dealers in the
farming localities.

Although it may not be feasible to have a regional research
station in every district, one or more research officers should be
located there to conduct adaptive research, particularly to
identify appropriate rice varieties and management practices
for the area. These officers would cooperate with the extension
field personnel distributed among the farming localities in
carrying on additional verification trials and on-farm demonstrations. Furthermore, they would link the principal national
rice research stations with the extension personnel residing in
the farming localities.

Each district requires an extension administrative office to
provide the field staff in the farming localities with periodic
training, with visual aid materials, with the most recent
information on rice research, and with ready access to
specialists. The office of the extension personnel and of the
research staff could well be in the same building, thus
encouraging close cooperation and coordination. In small
districts only one research officer and one extension worker
might be required. In such circumstances, however, both
should be easily able to call specialists into the district

whenever difficult problems arise on farmers' rice fields. Furthermore, they should be able to attend training courses from time to time to keep abreast of new developments.

The credit facilities in the farming localities can meet the production credit requirements of farmers only if they can rediscount their loans to farmers with larger banks or credit agencies. Thus, each district should contain at least one service unit with the resources to back up lending operations of small credit offices situated in the market centers of the farming localities. In most countries, even the district banks or credit offices are supported by a national banking structure.

Suitable roads must connect the market centers in the farming localities with the district center, where the larger markets exist. The roads should be adequate to take care of all types of vehicular traffic, including four-wheeled trucks, buses, and automobiles. Besides roads, it is advantageous to have good telephone and mail services.

All the activities taking place in the localities and districts should be coordinated with the programs of village, municipal, and provincial (or state) officials. The objective would be to strengthen and add to programs already in existence rather than to supplant them or to compete with them.

Deciding Where To Put the Emphasis

Selecting the Land Areas for Attention

Few countries have enough skilled manpower or funds to mount a nationwide accelerated rice production program all at once. If scarce resources are spread too thin, the program will have little impact on production. Most development authorities agree that any program to improve crop yields greatly should be started first in those localities and districts where the natural advantages are most abundant. For rice production, the chances of success are the greatest where irrigation facilities already exist, where the topography is level, and where the soils are heavy textured with rather low permeability to water.

In addition to a favorable natural setting, areas with a strong potential for initial success also would possess such organizational and infrastructural elements as a well set up extension service, adequate farm-to-market roads, sources of rural credit, and connections with the national economy. Such conditions can lead to early success at minimum cost.

All the less developed rice-growing countries contain land that has a potential for increased rice yields but that requires the investment of large amounts of money and manpower if high and stable production is to be obtained. The element most commonly lacking in these regions is water control. Irrigation and flood control projects are so expensive that developing countries usually have to seek foreign aid to undertake them. Nevertheless, rice production in the absence of water control is risky. Irrigation systems, if properly designed and operated, in the long run guarantee high returns in the form of increased rice yields.

There are also parts of any rice-growing country where the potential for increasing rice production is so low that no funds should be invested in rice development projects. Hilly, rocky, or extremely sandy areas are unfit for rice growing. Often, however, they are well adapted for upland crops, for forestry, and for livestock enterprises.

If a reliable inventory of a country's land and water resources has not been made, it is advisable to conduct such a survey to identify the areas that have a good potential for rice production.

Adopting Practices and Policies from Abroad

The most difficult decision for government administrators in the low-income countries is how to allocate scarce resources for development. Among the questions to be answered in planning for an expanded rice production program is which practices and policies of the more industrialized and hence more affluent nations should be adopted—and which should not—by countries that have an abundant labor supply but limited capital resources.

Mechanization. In the United States, for example, the rice industry is fully mechanized. Operations from land preparation to harvest are done with heavy machinery or by airplane. Indeed, many Western agriculturalists believe the only cure for low rice yields in the tropics is the adoption of large-scale mechanized farming methods. This is not so; time and again it has been proved that there is no "economy of scale" in lowland rice production. In fact, small farms have higher yields on the average than large ones, mainly because they are managed more intensively.

Probably only where extensive new areas are opened to rice culture could large-scale mechanized rice farming be practiced. Even then, great caution would be needed. Small pilot projects should be tried first to determine the basic suitability of the area for rice growing. Care needs to be exercised to avoid overinvestment in equipment. Problems of transportation, marketing, milling, and storage should be analyzed thoroughly. Numerous large-scale projects have failed in the tropics merely because administrators decided that what worked in the United States could be transferred directly to the tropics.

In Japan, South Korea, and Taiwan, where rice farms are still small, there has been a great increase in recent years in the number of power tillers and in transplanting and harvesting equipment. This expansion in mechanization was triggered largely by industry's demand for labor, which created a shortage of workers in rural areas.

Undoubtedly, as off-farm employment opportunities increase in the less developed nations, the mechanization of small rice farms will be justified. At present, power tillers tend to be unprofitable unless at least 10 hectares of rice are being cultivated, either on a single farm or by contract work. Furthermore, any program for expanding mechanization should include a sufficient supply of spare parts and adequate repair services.

Most rice farmers in Japan, South Korea, and Taiwan use mechanical threshing equipment, often powered by electricity. In countries where rural electrification is not widespread, threshing will continue to be done chiefly by hand or with small portable threshers powered by gasoline engines. Several farmers can share in purchasing a small power thresher, thereby conserving time and labor.

It is doubtful that mechanical transplanters will prove profitable in the labor-surplus nations for some time to come.

However, as chemical weed control methods become less expensive, more and more rice is likely to be direct-seeded rather than transplanted.

The developed countries generally use modern rice mills. Such facilities, when operated at full capacity, tend to be more efficient than the traditional rice mills. Nevertheless, their wide introduction into less developed countries may not be the wisest policy. The bases for deciding whether to invest in large modern mills are volume of rice to be handled, the alternative opportunities for using the unskilled labor force released by modernization, and the availability of capital to invest in rice mills as compared with other capital needs. In free economies every encouragement should be given for private capital to be invested in the rice-milling industry. Japan and Thailand present good examples of a successful milling industry run by private enterprise.

Irrigation. Japan, South Korea, and Taiwan irrigate most of their rice. In fact, all countries with average rice yields of over 4 t/ha irrigate from 80 to 100 percent of their rice crop. Properly designed and well-managed irrigation systems are a good investment, for in the absence of water control, the yield increases expected from the use of modern rice varieties and of fertilizer may not be realized. If the less developed countries are to remain self-sufficient in rice, they will find it necessary to put more of their rice land under year-round irrigation. Recent studies in Thailand and the Philippines showed that satisfactory net profits could be obtained from irrigated rice. In well-managed rice fields, the return above variable costs ranged from US$250 to US$600 per hectare, which is twice the level obtained from rainfed rice farms in the same areas.

The use of fertilizer. The more advanced rice-growing countries use 5 to 10 times as much fertilizer as the less developed countries. Although it is not economical to apply sufficient fertilizer to obtain the absolute maximum yield, because fertilizer response curves follow the law of diminishing returns, nevertheless appropriate input pays high dividends. Hundreds of fertilizer trials conducted on both farmers' fields and experimental farms show that in South and Southeast Asia, almost without exception, modern rice varieties give an economic yield response to at least 60 kg/ha of nitrogen in the wet season and to 100 to 120 kg/ha in the dry season, assuming that there is good water and weed control and no serious damage from insects, rats, and birds. When phosphorus, potassium, and zinc are limiting, those elements must be added. The only way to be certain of what nutrients should be applied is to conduct on-farm trials. The importance of this practice cannot be overemphasized. In some areas phosphorus or zinc is so limiting that the application of nitrogen alone produces no grain yield response.

Without question, if average national yields are to be increased, fertilizer must be made available to farmers at a price they can afford. It is important to keep the price ratio of nitrogen to paddy below 2.5 to encourage farmers to purchase enough fertilizer.

The more developed countries, such as Japan, South Korea, and Taiwan, manufacture their own fertilizer. Many nations manufacture part of their fertilizer needs and import the remainder. Decisions on whether or not to construct fertilizer factories—usually a matter for government action—depend on the size of the agricultural area, on its natural resources, and on the availability of capital funds.

Pest control methods. Rice farmers in Japan, Taiwan, and to some extent in South Korea, use vast amounts of insecticides and herbicides. Many rice specialists believe that the quantities used are excessive. Certainly, the less affluent nations should adopt less expensive methods of pest control. Above all, every effort should be made to develop rice varieties that are resistant to insect attack. This will continue to be the most economical and most effective way of reducing damage by insects.

In addition, the application of integrated pest control methods should be encouraged. Often a combination of the use of resistant varieties, optimum planting time, proper plant spacing, and other beneficial cultural practices, can keep insect populations under reasonable control. It should be remembered that in the absence of high insect populations, the use of insecticides is never profitable. A pest outbreak warning system encourages farmers to use pesticides only when necessary.

In labor-surplus economies, weed control in transplanted rice can be done by hand. However, administrators should carefully watch developments in weed control, since new herbicides may come on the market that are more economical than hand weeding.

In Japan and in several other countries, fungicides are used to control the rice blast disease. It is not recommended that less developed countries apply any chemicals for disease control in rice. The use of resistant varieties should provide adequate control provided that each country maintains a vigorous testing program so that as new physiologic races or strains of a disease appear, another rice variety will be ready to be substituted for the one that became susceptible.

Price support policies. High minimum support prices for rice in Japan, South Korea, and Taiwan have helped those nations attain rice self-sufficiency. Less industrialized countries, however, cannot afford to finance such expensive programs (Japan's support price, for instance, is several times the world price). Nevertheless, minimum support prices for rice should be maintained in all countries wishing to increase their rice production. Although authorities disagree on the proper level of support, most feel that the guaranteed prices should not be far from the world price. In years of crop surplus, the government can buy the nation's excess rice, accumulate buffer stocks, and arrange for some exports. In years of a deficit, prices automatically will rise above the support price, and the government will not have to purchase any rice from farmers.

Farmers' organizations. The traditional rice farmer has shown often and convincingly that he is willing to try innovations in technology if he believes they are feasible and profitable in his environment. Yet yield levels on farmers' fields have not risen to the extent possible even with current technology. Among the institutional (as opposed to technological) constraints to increased production, inadequate diffusion of knowledge ranks high in significance. Although a well-trained and adequately supported extension staff is essential, some sort of farmers' organization must also exist so that extension personnel can meet with rice growers as a group and so that the individual farmer has a better chance to become personally involved in the improvement of life in his community.

Farmers' organizations range from those as complete as the type found on Taiwan, which take care of marketing and extension activities, supply inputs at reduced prices, and furnish several other services, to simple cooperatives that do little more than provide fertilizer and seed at decreased prices and extend credit to member farmers for the purchase of those necessities.

The irrigation associations of Taiwan are particularly successful. Their organizational and operational patterns could well be duplicated in many less developed rice-growing countries. The most important benefit derived from having farmers belong to an irrigation association is that by participating in management decisions they see more clearly the water needs of the area and cooperate more willingly in promoting equitable water distribution throughout the command area.

When the users of irrigation water are unorganized, on the other hand, the individual farmer tends to grab whatever water comes down the ditch, without due regard for others. The farmer near the start of an irrigation canal gets more water than he needs, while the farmer near the end of the command area cannot get enough to supply his rice crop properly. Irrigation associations composed alike of farmers and of the field staff of the water system generally can solve such problems.

The communes of China, although subject to considerable government control and regulation, bind the farmers together as a productive unit. Goals are set and, to a degree, remuneration is determined by productivity. Each person has his assignment and seems to take pride in fulfilling it adequately. This highly controlled system of social organization would not be workable in many other cultures. It is successful in China largely because most decisions regarding crop production are made within the commune itself. Thus the commune can be a practical and realistic response to national policy rather than an automatic reflection of it.

Agricultural cooperatives have met with varying success in different countries. When properly organized and managed by competent people, they can contribute importantly to increased crop yields and to rural development generally. Too often the lack of success of cooperatives has been due to poor management. Most countries need training courses for managers of cooperatives.

A lesson to be learned from the more affluent nations is that nonpolitical farmers' associations led by competent and well-trained people who are dedicated to the lot of the small farmer contribute greatly to the advancement of agriculture. It is doubtful that the excellent progress in raising rice yields in South Korea and Taiwan would have been possible without the *Saemaul Undong* in the former and the Farmers' Associations in the latter. In fact, one can think of no advanced agricultural country that does not have organizations in which farmers are involved in the improvement and promotion of the crop or the animals they raise.

The less developed rice-growing countries sorely need to involve their farmers in rural development programs at the local level. Too commonly today, the poorer rice farmer is at the mercy of the moneylender and the unscrupulous politician and is not given an opportunity to become an active citizen involved in community development. The farmer is the key man in any crop production program: unless he participates in the planning as well as the day-to-day operations, a stepped-up rice production program is likely to move very slowly.

The Importance of Rice Research

Insufficient research on new agricultural techniques and materials often is a major cause of slow progress in crop production and agricultural development (a second major cause may be constraints that affect the farmers' willingness or ability to achieve the yield potential of the crop on his own farm). Although dramatic progress has been made in the last two decades in breeding improved rice varieties and in finding the best management methods for them, much significant research remains to be done. The barriers to further yield increases under less than ideal environmental conditions cannot be reduced without widespread research at both the national and the international level.

In spite of the great progress in rice research made by three international organizations—IRRI, CIAT, and IITA—during the past 15 years, national rice research programs continue to be essential for advancing knowledge about rice within each country. The climate, the soils, and the array of insect pests and diseases vary from region to region. Research to develop appropriate varieties and management methods for the varied environments should be conducted where the problems are. Even in the Philippines, for example, the varieties and the management methods developed at IRRI have to be tested in other parts of the country before they can be recommended to farmers.

On the other hand, no nation's research program should be conducted in isolation; many results obtained in one country are applicable in others. For instance, the varieties and cultural practices that are suitable for the Chao Phraya river basin in Thailand should work well in many parts of the Irrawaddy basin in Burma. Likewise, techniques developed in the Punjab of India can be used in many parts of Pakistan.

In spite of the transferability of rice varieties and management practices, there are numerous situations that differ sufficiently to make it necessary to test research results and materials widely before firm recommendations can be made. Therefore, systems for testing genetic materials and agronomic practices should be developed and maintained. Fortunately, national rice research programs and international organizations that work with rice are cooperating in such systems. The international rice nurseries coordinated by IRRI provide an excellent opportunity for all rice-growing countries, whether large or small, to test the most complete collection of genetic materials and the most advanced cultural methods that rice scientists have discovered. No tropical or sub-tropical rice-growing country should fail to participate in these international programs.

Prospects for Increasing Rice Yields in the Tropics

Irrigated Rice

There are abundant opportunities for increasing tropical rice yields on irrigated land, using methods and techniques now at hand. Modern varieties that are early to medium-early, photoperiod-insensitive, short-statured, heavy-tillering, fertilizer-responsive, and resistant to attack by most of the major insect pests and diseases have opened new vistas for yield increases where good water control exists. Since the late 1950s, the yield potential of lowland flooded rice has doubled. This great rise in yield, even under conditions of good water control, can be realized only if adequate fertilizer is applied, weeds are controlled, and severe damage from insects and other pests such as rodents and birds is prevented.

In any country, if the yield of irrigated rice is less than 4 t/ha, the most obvious actions required to raise yields are (1) to see that irrigation and flood control systems are well maintained and supervised, (2) to be certain that adequate fertilizer is available at a reasonable cost in relation to the price of rice, and (3) to take steps to provide adequate pest control at minimum cost.

In a recent study, agricultural economists at IRRI estimated that the average yield of irrigated rice in South and Southeast Asia is currently about 3.0 t/ha. They predict that by the 1990s it will be 4.1 t/ha. Theoretically, when management levels on farmers' fields equal those of currently conducted on-farm trials, per hectare yields under irrigated conditions could be raised another ton or more (see Table 3). At the least, administrators should feel confident that yields of over 4 t/ha on irrigated land are obtainable wherever good water control is possible.

Rainfed Rice

In contrast to the relatively simple steps that need to be taken to raise yields of irrigated rice, improving the production of rainfed paddy, where too little or too much water frequently reduces yield, is more complex. Because the returns are less certain, the decisions regarding the allocation of resources become more difficult. The magnitude of the problem is brought out by the fact that of the 83 million hectares of rice in South and Southeast Asia alone, half is classified as rainfed paddy.

On sloping land that has been terraced or on low-lying areas where water depths range between 15 and 100 centimeters during the rainy season, modern varieties usually are not suitable, although rice scientists are developing varieties that can tolerate deeper water and that have a greater drought tolerance. Administrators of rice production programs should see that seed of the more promising varieties is multiplied and distributed as soon as they have been proved suitable in adaptive research trials.

Because of the wide variation among countries in natural, social, economic, and political environments, it is impossible to make recommendations that can be applied generally. By and large, though, it would seem wise for nations with large areas of rainfed rice to make major investments in irrigation systems and, where needed, in flood control projects. Furthermore, research on rainfed rice should continue and be intensively pursued. Farmers should be encouraged to grow the best varieties available, which may be those that are of medium height (to tolerate deeper water) and early maturing (to escape drought). Moderate amounts of fertilizer should be applied, hand weeding (instead of chemical herbicides) should be employed, and pesticides should be used sparingly. This sort of strategy is aimed at keeping production costs low while raising yields above those now being obtained.

The greatest investment in rainfed rice should be made in low-lying and relatively level flood plains and river deltas. If rainfall is normal, yields in such areas often approach those of irrigated rice. Terraced rainfed paddy, on the other hand, is subject to severe moisture deficiency when periods of drought occur.

As is shown in Table 3, maximum yield trials on farmers' fields in 11 countries of South and Southeast Asia indicate that theoretically it is possible to obtain average yields on rainfed paddies ranging between 2.5 and 4.0 t/ha. This is well above current average yields for rainfed rice, which are less than 1.5 t/ha.

Maintaining the Pace Once It Is Set

Too frequently, the pace of rice production programs slackens following an initial success. Personnel and funds often are diverted to other projects, which indeed may be important to total agricultural development but which nevertheless do not directly increase rice production. In countries, or areas within countries, where rice is the dominant food crop and source of farm income, agricultural officers and planners must realize that to be successful, any accelerated rice production program has to have continuous support. If rice research and extension programs are curtailed, or if such development projects as irrigation and flood control are slowed down or stopped, rice yields soon will level off.

Often both farmers and extension staff feel that if a locality has been able to double its rice yields to, say, 2.5 t/ha, further increases are unlikely. The fact is that in areas of good water control, the yields of lowland rice could be doubled again. South Korea is an example (see chapter 5) of a country that did not content itself with initially increased production. Observing in the late 1960s that its average rice yields had leveled off at about 4.5 t/ha, South Korea mounted an intensive rice research and extension program tied to a nationwide community development effort and was able to add more than a ton to its national average rice yield.

The press has used such terms as "miracle rice" and "Green Revolution" in reporting the advances made in increasing rice yields. Although such phrases arouse popular interest in agricultural research, they are unrealistic. Crash rice production campaigns have served to get a program off to a quick start, but the real basis for progress is more evolutionary than revolutionary. Sustained support of research and extension programs, steady expansion of irrigation systems, the provision of fertilizer at reasonable prices, and the formation of farmers' organizations are examples of important actions that form the core of a successful national rice production program.

It is to be remembered that the countries that have achieved high average rice yields did so through careful planning and concerted action. There was always present a national will to reach the desired goal. None of those successes came about by chance or with ease. Although the point has been made repeatedly that rice yields can be doubled in any country now producing less than 2 t/ha, it must be borne in mind that there are limits to land area, to solar energy, and to the world's nonrenewable resources. Thus, there is a limit in the long run to the amount of rice and of other food crops that can be produced and, accordingly, to the number of people the earth can support.

The significant reductions in population growth rate have occurred largely in the more developed and affluent countries, where crop yields are high. Nevertheless, it is encouraging that many of the overpopulated, less developed nations have mounted sizable population control programs. Certainly, the impact of those programs should become visible in the next decade or two. In the meantime, there is ample opportunity for the major rice-growing countries to produce enough rice to feed their ever-increasing inhabitants well into the next century.

Appendix:
Where and How
To Get Assistance

Often countries need both technical and financial assistance to mount an accelerated rice production program. The successful programs described in chapter 5 received substantial aid from foreign sources in the course of development. This appendix describes major international organizations that offer the less developed countries financial and technical assistance for the improvement of their rice production efforts.

International Agricultural Research Centers

There is a worldwide network of 11 international agricultural research centers, three of which have rice research and training programs.

International Rice Research Institute

The most important source of technical assistance to the rice-growing countries is IRRI. Located in Los Baños (about 65 kilometers from Manila) in the Philippines, its environment is typical of much of the humid tropics; thus, the results obtained on its experimental fields have wide applicability in tropical regions around the world.

Foremost among IRRI's activities is a sound, comprehensive rice research program embracing all aspects of plant and soil science, including agricultural engineering, agricultural economics, and cropping systems. This program is conducted by an experienced, competent senior staff composed of some 40 scientists from nine nations. In addition, about 20 senior scientists are assigned to cooperative national programs. In 1977 these scientists were located in Indonesia, Sri Lanka, Thailand, the Philippines (beyond IRRI headquarters), Bangladesh, and Pakistan.

IRRI's international activities are designed to develop links with national programs, to help strengthen national capabilities, and to facilitate collaboration with national scientists in solving major rice production problems. To accomplish those objectives several approaches are used.

IRRI scientists are available to travel to any rice-growing country in the world to assist in identifying problems. IRRI conducts cooperative country projects and collaborates in regional projects. It organizes international rice network investigations and testing, has a large training program, and annually holds 5 to 10 workshops, conferences, or symposia to which scientists and administrators from around the world are invited.

The International Rice Testing Program is the most comprehensive global cooperative venture of IRRI. It is an extension of IRRI's Genetic Evaluation Unit, which uses the "team" approach in a greatly intensified rice genetics and breeding program designed to produce varieties that can tolerate many of the pest and environmental stresses that afflict the rice plant. To screen the thousands of varieties and genetic lines from IRRI's breeding program, from its germ plasm collection of over 45,000 entries, and from national rice breeding programs, an extensive worldwide testing system has been developed. In 1977 the network included 14 different types of rice nurseries ranging from observational and yield trial nurseries to those that screen for tolerance to specific pests and diseases, to extremes of temperature, and to adverse soil conditions. The magnitude of the program is revealed by the fact that a total of 99,000 seed packages were delivered to cooperating programs in 1976. Nearly 50 percent of the entries in these nurseries were nominations from cooperating national programs; the remainder came from IRRI's breeding program and from its germ plasm collection. Thus, much of the diversity in the world's rice germ plasm finds its way into the International Rice Testing Program.

IRRI's training program is available to interested agencies throughout the world. The program can be divided into three categories: (1) research training for promising young scientists, including postdoctoral fellows (if desired, qualified candidates may pursue graduate study leading to an M.S. or a Ph.D. degree at the adjacent College of Agriculture of the University of the Philippines); (2) a 6-month rice production training course and a similar one in cropping systems (both of which are practical courses designed primarily for extension workers, field technicians, and farm managers); and (3) short courses to accommodate cooperators in the international rice testing programs or other abbreviated courses, lasting 2 to 4 weeks, to assist organizations that wish to train groups of employees who cannot be spared for longer periods. Thousands of people have been trained by IRRI since this program started in 1964. In 1976 alone, 244 research fellows, scholars, and trainees from 27 nations received some type of training at IRRI.

IRRI produces an annual report that gives considerable detail on its research findings and serves as a valuable source of information, particularly to rice scientists located in isolated research stations with inadequate libraries.

The IRRI Reporter is a 4- to 8-page quarterly publication featuring highlights of IRRI's research program. It is available free of charge to interested persons.

IRRI also publishes the *International Rice Research Newsletter,* which reports what science is doing to increase rice production. All scientists working with rice are invited to contribute concise summaries of significant rice research. The *Newsletter* is published several times a year and appears to be the best available single source of current global research on rice.

IRRI prints special brochures and leaflets about such activities as its training program and international rice-testing program. It publishes technical reports written by its scientists and recently started a new publication, the *IRRI Research Paper Series,* which features articles by senior scientists.

Also published by IRRI are the proceedings of its symposia, workshops, and conferences. These are available to the participants and, on request, to institutional libraries.

Colored slides and several other visual aids prepared at IRRI, largely depicting the major research achievements in modern rice research, can be obtained by appropriate organizations on request. Agencies wishing to receive IRRI publications or visual aids may write to the Office of Information Services. Other initial requests for assistance should be sent to the director general, who will refer them to the appropriate department or program.

The International Rice Research Institute
P.O. Box 933
Manila, Philippines

Centro Internacional de Agricultura Tropical

CIAT (International Center for Tropical Agriculture), located in Colombia, does not consider rice research to be a major part of its program. Nevertheless, its well-run, productive rice program, conducted in close cooperation with IRRI, has been able, because of CIAT's geographic location, to make significant contributions to rice production in Latin America.

CIAT has a small rice research staff composed of breeders, agronomists, and plant pathologists. It conducts training courses in rice production and accepts rice research scholars for extended stays to work under senior scientists. Its rice breeding program develops varieties particularly adapted to Latin American conditions. CIAT likewise runs cooperative trials with national programs, testing varieties and genetic lines, and conducts field experiments with fertilizers, herbicides, and cultural practices such as seeding rates, comparisons between transplanting and direct-seeding, etc.

Countries needing assistance in rice production programs will find knowledgeable and cooperative people at CIAT who have access to the newest genetic materials and experience in modern rice management practices.

CIAT publishes its research findings in an annual report. The rice improvement section is available, as a separate document, to interested persons upon application.

Centro Internacional de Agricultura Tropical
Apartado 67-13
Cali, Colombia

International Institute of Tropical Agriculture

IITA conducts research on rice but does not consider this work as one of its major activities. Nevertheless, its location in Nigeria and its proximity to other African countries enable it to render effective help to African nations in expanding their rice industries.

IITA has an active research program with a capable staff of plant breeders, agronomists, entomologists, plant pathologists, economists, and agricultural engineers, most of whom devote only part of their time to rice research.

There is a training program for research scholars and occasionally a rice production course for extension workers. IITA accepts postdoctoral fellows, who may work for a year or two with the senior scientists, gaining experience in their chosen specialty but working with rice.

The IITA *Cereal Improvement Program Annual Report* may be obtained by writing to the office of communications and information.

International Institute of Tropical Agriculture
PMB 5320
Ibadan, Nigeria

Private Foundations and Other Nongovernmental Organizations

Although there are numerous foundations and nonprofit organizations that support agricultural research, education, and development, only those with substantial worldwide activities are mentioned
here.

Rockefeller Foundation

The Rockefeller Foundation (with home offices in New York) is
spending over US$6 million per year in its agricultural sciences
program. Its funds go largely to research and educational institutions
for library books, laboratory equipment, and the granting of
fellowships for advanced study. In addition, the foundation places
scientists at various institutions around the world to strengthen
research and educational programs. In 1976 there were 29 long-term
visiting scientists from the United States stationed abroad. The foundation also provides short-term consultants in disciplines where it
has special competence.

The principal interest of the foundation's agricultural program is
to strengthen research and educational institutions in the areas of
crop and animal production and human nutrition.

The Rockefeller Foundation
1133 Avenue of the Americas
New York, N.Y. 10036, U.S.A.

Ford Foundation

The agricultural activities of the Ford Foundation (which is based
in New York) are directed toward the support of research and
education in the less developed countries. In 1976 the Ford
Foundation made grants for agricultural purposes of more than US$8
million. Its current agricultural program provides support for studies
of (1) crop production technology, (2) socioeconomic factors involved
in increasing food production, (3) the management of ecological
variables affecting food producing systems, and (4) means of
improving the social and economic status of rural populations. The
foundation may support studies that use only indigenous staff, or it
may supply visiting scientists or consultants. In addition, it supplies

training where needed to strengthen a special project or program.

Representatives of the Ford Foundation are located in Bangladesh, India, Pakistan, Thailand, the Philippines, Kenya, Tunisia, Nigeria, Egypt, Brazil, Peru, Colombia, and Mexico. Nations seeking assistance should contact the nearest representative. The headquarters in New York can provide addresses of its overseas offices.

Ford Foundation
320 East 43rd St.
New York, N.Y. 10017, U.S.A.

Agricultural Development Council

The primary aim of the Agricultural Development Council is to strengthen training, research, and educational activities in agricultural economics and rural sociology mostly in the less developed countries of South and Southeast Asia. Among its activities are fellowships for foreign study and the assigning of specialists to educational and research organizations. It also supports country studies in agricultural development and prepares valuable books and pamphlets on that subject.

Agricultural Development Council
630 Fifth Avenue
New York, N.Y. 10020, U.S.A.

International Agricultural Development Service

IADS is a new organization that provides the less developed countries of the world with assistance in their agricultural development programs, through whatever means seem most appropriate. Primarily its efforts are aimed at improving national research and extension programs that are particularly related to increased agricultural productivity. Teams of specialists are assembled, on request, and sent to countries for short periods to identify bottlenecks in agricultural development and to suggest remedies.

IADS is not a fund-granting organization, although it is willing to consult with governments free of charge. According to the first annual report of IADS, it provides services "at cost to individual countries when (1) national authorities so request, (2) IADS's capabilities are applicable, (3) there is an agreed-upon program of work, (4) suitable financing is available, and (5) IADS trustees approve the arrangements."

IADS lists five ways in which it best can assist the less developed countries: (1) by analyzing development opportunities, (2) by examining national agricultural research and training programs and recommending improvements, (3) by helping plan and implement crop and animal production programs, (4) by assisting in planning and implementing agricultural development programs in specific areas, and (5) by strengthening colleges and schools of agriculture.

International Agricultural Development Service
1133 Avenue of the Americas
New York, N.Y. 10036, U.S.A.

Major International Lending Agencies

Included in this section are the few major international and regional organizations that do not customarily make grants but, rather, confine their activities to lending money to governments.

International Bank for Reconstruction and Development

IBRD and its affiliate the International Development Association (IDA) are frequently referred to as the World Bank; they have the largest lending program for agriculture in the world. IDA makes concessionary (low-interest or long-term) loans to poorer nations that cannot afford to borrow money from IBRD on conventional terms.

World Bank loans are extended to member countries (about 130 in 1977) for many kinds of development projects covering both industry and agriculture. Approximately 25 percent of all funds lent by IBRD/IDA are for agricultural projects. In 1976 total lending for agricultural projects and programs by both IBRD and IDA amounted to around US$6600 million, of which about 60 percent was actually disbursed in that year. These loans were distributed among 90 countries.

Lately the World Bank has attempted to redesign its lending policies to benefit larger numbers of the rural poor by focusing on projects in area development, settlement, irrigation, and land improvement. Most of the projects have an agricultural base often including irrigation systems, rural credit, fertilizer availability, etc. Installation and improvement of irrigation schemes are the activities that have been receiving the most support.

Although the World Bank is not primarily a grant-giving agency, it makes feasibility studies and surveys free of charge, when there is some likelihood of mutual interest between the country requesting a loan and the bank.

To help the less developed countries in preparing loan proposals, the bank has regional offices in Kenya for Eastern Africa, in the Ivory Coast for Western Africa, and in Thailand for Southeast Asia. In addition, the World Bank has missions in many less developed regions to assist governments in managing loans. In 1977, country missions were located in Afghanistan, Bangladesh, Cameroon, Colombia, Ethiopia, Ghana, India, Indonesia, Nepal, Nigeria, Pakistan, Sudan, Tanzania, Upper Volta, Venezuela, Zaire, and Zambia.

World Bank
1818 H Street, N.W.
Washington, D.C. 20433, U.S.A.

Asian Development Bank

The ADB, with headquarters in the Philippines, makes many loans for rice projects because of the great importance of rice in Asia. There are 40 member countries supporting the ADB, 26 of which are in Asia. Japan, which is the only fully developed Asian country, provides major support to the bank. The ADB lends only to the less developed member countries (the other countries being contributing members). During the past several years, the ADB has been lending some 26 member countries between US$200 million and US$250 million annually for agricultural projects. This is about one-fourth the amount provided yearly to the same countries by the World Bank.

Asian countries with severe agricultural development problems will find the ADB sympathetic to their needs. The bank is seeking well-organized proposals for loans that promise to have a significant impact on food production and economic development.

Asian Development Bank
P.O. Box 789
Manila, Philippines 2800

Inter-American Development Bank

The IDB has about 40 member countries, 24 of which are Latin American nations eligible to receive loans. The remainder are more affluent countries that contribute materially to the IDB's program.

In 1976, the IDB provided US$1500 million in loans, 28 percent of which were allocated to agricultural projects. Commanding its top interest are programs to accelerate production of specific crops, irrigation and drainage systems, multipurpose agricultural credit systems, land settlement and agrarian reform, and storage and marketing systems. In the past 15 years, about 30 percent of the IDB's loans have been used for irrigation projects, 30 percent for credit, and 20 percent for integrated development and agrarian reform. The balance of the funds was lent for projects in marketing, in research and extension, in animal health, and in fishing and forestry.

The IDB spends about US$7 million annually in what it terms "technical cooperation," which is directed primarily to feasibility studies of new loan applications. Generally these are considered as grants and do not need to be reimbursed.

Periodically, the IDB sends a programming mission to each borrowing member country to help identify important projects that might qualify for loans. For a proposal to mature into an actual loan often takes 3 years.

IDB maintains field offices in the capital cities of essentially all the Latin American nations that are eligible for loans. Therefore it is relatively easy for governments to keep in touch with the IDB.

Inter-American Development Bank
808 17th Street, N.W.
Washington, D.C. 20577, U.S.A.

African Development Bank

Located in the Ivory Coast, the African Development Bank has 36 member countries, all of them in Africa. It has had difficulty in raising capital; nevertheless, it has lent US$36 million for agricultural projects since the beginning of its operations in 1965. In addition to its conventional loans at normal interest rates, the African Development Bank extends credit on easy terms to the poorest countries of Africa. So far, about 40 percent of these credits have been for agricultural purposes.

African Development Bank
B. P. 1387
Abidjan, Ivory Coast

Multinational Organizations

Food and Agricultural Organization

The FAO is the world's largest agency devoted to the improvement of human nutrition and to increased food production. It is supported by 136 member countries. Its aim is to assure freedom from hunger for all of humanity.

Member countries contribute to the FAO budget in proportion to their gross national product. In addition, the FAO receives funds from other agencies, such as the United Nations Development Programme and the World Bank, for projects that it administers and conducts for them. Currently, the FAO spends about US$500 million annually on 1200 projects around the world.

One of the FAO's highly important contributions comes from its statistical office, which collects, analyzes, interprets, and disseminates information relating to nutrition, food, and agriculture. Its compilation of data on area, yield, and production of food crops is the most complete anywhere. Its predictions of future supplies and its analyses of the state of food and agriculture—information that is published periodically—are of great help to food-deficient nations in planning future programs.

The FAO conducts many workshops and conferences on major food crops and on some of the principal components of agricultural development.

FAO supplies experts to many less developed countries on long-term assignments. Of its 3000 professional employees, about 2000 are assigned to national programs or work in the FAO's various regional offices. Some 102 of the developing countries have a United Nations representative and staff that handle activities of the FAO and the United Nations Development Programme on a country-by-country basis. The country representatives can supply information about the FAO's varied and widespread programs to governments wishing assistance. For UN member countries that do not have representatives and that wish to seek aid, the FAO maintains four regional offices: in Ghana for Africa, in Thailand for Asia and the Far East, in Chile for Latin America, and in Egypt for the Near East.

FAO
Via delle Termi di Caracalla
00100 Rome, Italy

United Nations Development Programme

UNDP is the principal source of United Nations funds for technical assistance. It dispenses over US$300 million annually, of which about one-third goes to agricultural projects among 140 member countries of the United Nations.

The United Nations representative and his staff in each developing country periodically review the development plans of the country and with it agree on major areas of UNDP support. Within his budget, the representative has authority to approve projects up to US$250,000. Proposals involving larger funding are sent to headquarters in New York for review and approval.

Generally, UNDP assistance in agriculture is administered by the FAO field staff. There are field representatives handling UNDP and FAO activities in 108 of the less developed countries.

United Nations Development Programme
One United Nations Plaza
New York, N.Y. 10017, U.S.A.

European Development Fund

The EDF was created by the European Economic Community (EEC) in 1975 as a means of supplying financial and technical assistance to 52 African, Caribbean, and Pacific associated states. Most of the 52 countries are former colonies of the EEC members. The nine full members of the EEC are Belgium, Denmark, France, West Germany, Ireland, Italy, Luxembourg, the Netherlands, and the United Kingdom. These countries provide the budget for the EDF.

From the time that the EDF became fully operative (in April 1976) until February 1977, it had approved projects in rural production costing US$90 million, of which US$87 million was for grants and the remainder for loans. The terms of the loans are "soft," only 1 percent being charged annually, with 40 years for repayment.

EDF projects attempt to complement national efforts by the 52 associated states. The program emphasizes capital projects in rural development, programs to improve crop and animal production, technical cooperation programs in the areas of training and technological adaptation or innovation, and projects at the grassroots level for development in rural areas.

For 1976 through 1980, the EDF is budgeting US$3000 million for its entire program. The money is allocated to the 52 countries by the EDF, and a representative of the latter is assigned to each country. These EDF delegates help develop projects and oversee their execution.

Fonds Européen de Développement, Communautés Européenes
Rue de la Loi, 200
B1049 Brussels, Belgium

West African Rice Development Association

WARDA is a multinational rice research and development organization that has 14 West African countries as members. Its headquarters are in Liberia, and its governing council is composed of one representative from each member state. WARDA's purpose is to assist national governments in developing their rice production programs. This is accomplished largely by strengthening their rice research, assisting in the training of both research and extension personnel, and helping to prepare rice development projects. In addition, WARDA sponsors workshops, conferences, and symposia to which most of the rice scientists of West Africa are invited. This is a most important contribution—before WARDA was started in 1971 there was little communication between the anglophone and francophone countries in West Africa.

Although WARDA receives some support from each of its member countries, it is largely financed by multilateral and bilateral grants from various foreign assistance agencies. In 1977 WARDA's total budget was in excess of US$2 million.

WARDA conducts a rice production training course at the College of Agriculture of the University of Liberia. This annual course (lasting several months) is open to candidates from all member countries.

There is close cooperation among WARDA, IITA, and IRRI, which participate in one another's workshops and exchange experimental data and ideas. WARDA is an active cooperator in IRRI's International Rice Testing Program.

Besides conducting coordinated varietal trials in member countries, WARDA in 1977 was receiving bilateral grants for research on mangrove swamp rice in Sierra Leone, on deep-water rice in Mali; and on irrigated rice in Senegal.

WARDA's development office works with national governments in preparing development projects, mainly irrigation schemes. Its statistical office compiles and distributes detailed data on rice in the African countries.

WARDA maintains a small headquarters staff comprising the executive secretary and his deputy, a research coordinator, a development officer and several plant breeders and agronomists.

West African Rice Development Association
P.O. Box 1019
Monrovia, Liberia

National Foreign Assistance Programs

United States

The U.S. Agency for International Development (USAID) under the authority of the Department of State is responsible for the bilateral assistance program of the United States. In 1976, its financial aid to agriculture totaled US$582 million, of which US$116 million was provided in grants. The remainder was used for loans.

USAID's program includes (1) assistance in developing policy and planning, and in strengthening local institutions; (2) development and diffusion of new technology; (3) helping to assure the availability of agricultural inputs, such as fertilizers, pesticides, irrigation systems, farm machinery, and rural electrification; (4) assistance in developing infrastructure, such as rural roads, storage and marketing facilities, agricultural credit, and agricultural businesses.

USAID has missions in most of the less developed countries with which the United States has diplomatic relations. The field missions, in cooperation with the host countries, prepare proposals for USAID assistance that go through several stages to determine mutual interest. If there is general agreement along the way, a final project paper is prepared that usually is sent to Washington for formal approval.

Because the nations that have invited USAID to place missions in their countries are essentially the only ones that receive help, government agencies in those countries should start any inquiries regarding assistance with the local mission in the capital city.

U.S. Agency for International Development
Washington, D.C., 20523, U.S.A.

United Kingdom

The Ministry of Overseas Development (ODM) is responsible for all United Kingdom foreign aid, including that for agriculture. The ODM's aid program is extremely broad. It includes loans and grants for development projects, fellowships for training and direct technical assistance to research and extension programs in the less developed countries. Much of the ODM's aid is given to former British colonies and to the less affluent Commonwealth nations. The amount of assistance to renewable natural resources (agriculture, forestry, and fisheries) is about US$70 million yearly.

Specific project identification is the responsibility of the embassies or high commissions in the host countries. Preliminary projects are sent to the ODM for scrutiny. If there is sufficient interest at the home office, then an appraisal team of technical people may be sent to the field to examine the project's feasibility and chances of success. Government agencies wishing to explore possibilities of assistance in agricultural development should get in touch with the local embassy or high commission of the United Kingdom.

Ministry of Overseas Development
Eland House, Stag Place
London SW1E 5DH, United Kingdom

Federal Republic of Germany

Germany's foreign aid program is handled by several agencies. The Credit Bank for Reconstruction takes charge of capital loans. The German Agency for Technical Cooperation implements technical assistance. The German Foundation for International Development adminsters training.

In 1975 Germany's total foreign aid was US$1689 million. About US$542 million was used for loans to agriculture, of which two-thirds went toward irrigation projects, with lesser amounts for regional development, credit and marketing. In addition, Germany has a technical assistance program with about 40 percent of its activity allocated to agriculture. In 1976 there were 366 active projects, 185 of which involved German personnel assigned to foreign countries.

About 10 percent of the training awards are in agriculture. In 1975, 446 people from foreign countries received training in Germany.

The government agency in a less developed country that wishes to obtain financial or technical assistance from the Federal Republic of Germany should approach the local German embassy.

Credit Bank for Reconstruction (KfW)
Palmengartenstrasse 5-9
6 Frankfurt/Maine, Federal Republic of Germany

German Agency for Technical Cooperation (GTZ)
D-6236 Eschborn 1
Postfach 5180, Federal Republic of Germany

German Foundation for International Development (DSE)
Wielinger Strasse 52
8133 Feldafing, Federal Republic of Germany

Canada

Canada has two principal foreign assistance agencies, the Canadian International Development Agency (CIDA) and the International Development Research Centre (IDRC).

CIDA is a governmental organization directly responsible to the Secretary of State for External Affairs. Between 15 and 20 percent of its financial assistance goes to agriculture and amounts to around US$70 million annually. The more affluent among the less developed countries receive loans for major development projects at conventional interest rates. The poorer countries with low per capita incomes get concessional loans, which usually mean no interest and up to 50 years for repayment.

Canada has selected a group of about 25 low-income countries in which it conducts development programs. Missions are sent out to identify the projects within those programs. Canada provides individual project support to some 40 countries other than the 25 selected for long-term program assistance.

CIDA provides scholarships for training, gives technical assistance in making studies and surveys related to agricultural development, and participates in long-term technical assistance programs to build up research projects in the less developed countries. About 100 agricultural specialists are stationed abroad.

CIDA representatives stationed in Canadian embassies throughout the world can assist government agencies in applying for financial or technical aid.

IDRC was established by the Canadian Parliament in 1970 as an autonomous public corporation. Its policies are controlled by a 21-member Board of Governors, of which 11 are Canadian and the rest are from both developed and less developed nations. Approximately one third of IDRC's funds are directed to agricultural projects. In 1975-76 this amounted to US$12.6 million. IDRC provides grants, but no loans.

IDRC's interests with respect to rice are principally in multiple cropping and the improvement of postharvest handling of the rice crop. Its program almost entirely supports research. It assigns relatively few (about 15) experts to foreign countries; instead, the professional staff stationed in Ottawa travels frequently and extensively to provide guidance to the many projects IDRC supports.

IDRC maintains regional offices in Singapore, Colombia, Kenya, Senegal, and Egypt. However, governmental agencies desiring to explore possibilities of research assistance may direct their initial inquiries to the headquarters office in Canada.

Canadian International Development Agency
122 Bank St.
Ottawa, Canada K1A 0G4

International Development Research Centre
Box 8500
Ottawa, Canada K1G 3H9

The Netherlands

The administration of development assistance in the Netherlands is the responsibility of the minister for development cooperation in the Ministry of Foreign Affairs, who is assisted by the Directorate-General for International Cooperation within the same ministry. The Netherlands makes both loans and grants, and in 1975 expended US$62 million for agricultural projects. In the same year, over 600 experts served abroad on technical assistance projects in agriculture.

The nation is now directing its agricultural program toward the small farmer in the poorer countries. The research programs they support are designed to open new avenues of income for small-scale farmers. The Netherlands is interested also in assisting extension services and agricultural schools to make farmers aware of the possibilities for increased yield through the use of modern varieties, fertilizers, and irrigation and of the fact that those inputs are actually within their reach.

The country feels that it has special technical competence in land reclamation and drainage projects, plant breeding, plant protection, and rural extension.

It is interested in training young scientists and extension workers. This is done mostly at the International Agricultural Centre in Wageningen. In 1975 about 400 young professionals from foreign countries received training there.

Initial requests for assistance from the Netherlands should be sent through the embassy in each country, from which it will be forwarded to the Ministry of Foreign Affairs. The ministry, in turn, will pass the request on to the Directorate of Agricultural Assistance to Developing Countries.

Directorate of Agricultural Assistance to Developing Countries
(AHO) Bezuidenhoutseweg 73
The Hague, Netherlands

Japan

Japan's large foreign assistance program is confined chiefly to the less developed countries of Asia. The principal agricultural interests of the Japanese foreign aid program are rice research and rice irrigation and drainage projects.

Although the Japan International Cooperation Agency and the Overseas Economic Cooperation Fund handle most foreign assistance projects, the recommended initial approach by a government agency in a less developed country is via the Japanese embassy, which will pass the request on to the appropriate government unit in Japan.

In addition, the Agriculture, Forestry and Fisheries Research Council provides certain experts in accordance with requests from foreign countries.

Japan International Cooperation Agency
Shinjuku Mitsui Building, 21 Nishi, Shinjuku-ku
Tokyo, Japan

Overseas Economic Cooperation Fund
IINO Building, 1-1, Uchisaiwaicho 2-Chome, Chiyoda-ku
Tokyo 100, Japan

Agriculture, Forestry, and Fisheries Research Council
Ministry of Agriculture and Forestry
Kasumigaseki, Chiyoda-ku
Tokyo, Japan

Other National Foreign Assistance Programs

A number of other countries have smaller, though still substantial, foreign assistance programs that, for the sake of brevity, are not described here.

Belgium
Administration Générale de la Coopération au Développement
Ministère des Affaires Etrangères
Building "AG," Place du Champ de Mars 5
1050 Brussels, Belgium

Denmark
Danish International Development Agency
Amaliegade, 7
1256 Copenhagen, Denmark

France
Ministère de la Coopération
20 Rue Monsieur
75007 Paris, France

Norway
Norwegian Agency for International Development
Boks 8142, Oslo-Dep.
Oslo 1, Norway

Sweden
Swedish International Development Authority
Birger Jarlsgatan 61
S-10525 Stockholm, Sweden

Switzerland
Swiss Technical Cooperation
Eigerstr. 73
3003 Berne, Switzerland

Australia
Overseas Economic Relations Division
The Treasury
Canberra, A. C. T. 2600, Australia

Australian Development Assistance Bureau
P.O. Box 887
Canberra, A. C. T. 2601, Australia

New Zealand
External Aid, Ministry of Foreign Affairs
Private Bag
Wellington, New Zealand

Commercial Companies

Agribusiness concerns selling such materials as fertilizers, pesticides, and farm machinery often provide not only modest financial assistance to agricultural research institutions, but also conduct useful field demonstrations of their products. Furthermore, they offer excellent technical guidance for the use of their equipment or materials. Because it is to the advantage of the companies to have their products properly used, their advice is well worth following.

Glossary

Amylopectin. The kind of starch in the rice grain that tends to make the rice moist and sticky on cooking. The starch in glutinous rice is essentially 100 percent amylopectin.

Amylose. The kind of starch in the rice grain that tends to make the rice dry and fluffy on cooking.

Awn. A thin bristle-like structure protruding from the top of the lemma (one of the glumes) of bearded rice varieties.

Bacterial blight. A disease caused by the bacterium *Xanthomonas oryzae.* The symptom is long lesions, starting at the edges of the leaves and often resulting in dead tissue over a large portion of the leaf area of a rice stand. Usually the disease attacks the crop after flowering.

Basal application. Refers to chemical fertilizer that is applied to a rice crop just before seeding or transplanting.

Bearded. An awned rice variety (see *awn*).

Beri-beri. A human nutritional deficiency disease caused by a lack of thiamine (a component of the vitamin B complex) in the diet.

Brown rice. Dehulled rice from which the bran layer has not been removed. Approximately equal to 80 percent of the weight of paddy or rough rice.

Brown spot. A fungus disease of rice caused by *Helminthosporium oryzae.* The most typical symptom is the appearance of numerous oval brown spots on the leaves. The disease can also attack the glumes.

Bund. The dike or raised border surrounding a rice paddy that makes it possible to hold rain water or irrigation water on the surface of the ground.

Deep-water or *floating rice.* Rice that can be grown at water depths of from 1 to 6 meters. Deep-water rice varieties are different genetically from normal rice in that they have the ability to elongate rapidly when flood waters rise. In addition, they put out adventitious roots at the submerged nodes.

Denitrification. The biological breakdown of oxidized forms of nitrogen (nitrite, nitrate or nitrous oxide), resulting in the evolution of nitrogen gas that is lost to the atmosphere.

Endosperm. The starchy inner portion of the rice grain that remains after the hull, bran, and germ have been removed by milling.

Glumes. The outer hard covering of the rice grain that, botanically speaking, consists of the lemma and palea. Also called hulls.

Grassy stunt. A virus disease of rice causing severe stunting, excessive tillering, erect growth. The leaves turn pale green or yellow and are covered with rusty spots or blotches.

Harvest index. The ratio of grain weight to total above-ground dry weight. Thus if the weight of grain is half that of the total weight of the plant, the harvest index is 0.5, which corresponds to a grain/straw ratio of 1.0.

Head rice. Unbroken milled rice.

Hull or *husk.* The outer, inedible portion of the rice grain; the glumes.

Kneeing ability. The capacity of the upper stems and leaves of floating or deep-water rice to become erect after being in a horizontal position on the surface of rising flood waters. It is a geotropic response and a desirable characteristic.

Ligule. A small, papery, triangular structure at the base of the leaf blade of the rice plant.

Lodging. The falling over of the rice plant. It occurs around flowering time in tall varieties that are grown under moderate to high soil fertility conditions.

Lowland rice. Irrigated rice as well as rainfed paddy when grown on flooded fields.

Nitrification. The biological transformation in the soils of ammonia nitrogen to oxidized forms such as nitrite, nitrate, or nitrous oxide.

Paddy. Whole grain rice. Synonymous with rough rice (unhulled rice), approximately equal to brown rice multiplied by a factor of 1.25. Used also for a bunded plot of land on which rice is grown.

Panicle. The portion of the rice plant that bears the seeds. The seed head.

Photoperiod-sensitive varieties. Varieties of rice that will not flower until short day-lengths, usually less than 13 hours, occur. Conversely, photoperiod-insensitive varieties tend to be uniform in crop duration regardless of the length of day, especially under tropical conditions where temperatures are relatively stable.

Ragged stunt. A new (1976) virus disease of rice in Asia characterized by stunting and wavy, ragged edges of the leaves. There is little panicle exertion, and few grains are formed. The stunting is not accompanied by severe leaf discoloration as in the tungro disease or the grassy stunt virus disease.

Rainfed paddy. Rice that is not irrigated but is grown on leveled paddies that are bunded or diked to allow an accumulation of flood water on the surface during heavy rains.

Rice. The crop in general, the whole grain as harvested, and the milled edible product. All rice yields in this book are expressed in terms of paddy or rough rice.

Rice blast disease. A ubiquitous rice disease caused by a fungus (*Pyricularia oryzae*) that produces spots or lesions on leaves, nodes, and panicles.

Rice variety. A kind of rice that has been bred or selected, that is genetically uniform, and that breeds true from seed. A variety bred in a crossing program is synonymous with "cultivar." In many countries, varieties are named and released officially. Previous to release they may be referred to as "genetic lines."

Sheath blight. A fungus disease of rice caused by *Corticum sasakii.* Early symptoms are ellipsoid or ovoid greenish-gray spots on the leaf sheath that later may enlarge to be 2 or 3 centimeters in length.

Systemic insecticide. An insecticide that is absorbed by the plant through the roots or through the leaves and that kills insects that feed on the plant tissues. In flooded rice, systemic insecticides may be applied in granular form to the flood water or placed in capsules in the root zone.

Tiller. Any of the extra stems or culms in a rice plant that arise from its base. Varieties that have the genetic capacity to put out many stems are referred to as "heavy-tillering" varieties.

Ton. 1000 kilograms (2204 pounds).

Topdressing. An application of fertilizer that is applied to the soil or water surface after the crop is well established, usually at the panicle initiation stage about three weeks before flowering.

Tungro disease. A virus disease of rice causing stunting and a yellow to orange discoloration of the leaves. The disease is called *penyakit merah* in Malaysia and the *mentek* disease in Indonesia.

Turn-around time. The number of days between the harvesting of one crop and the planting of the next crop.

Upland rice. Nonbunded, direct-sown rice grown in a manner similar to that of wheat or other small grains.

Wooden dunnage. The low wooden platforms, with airspace beneath, on which sacks of rice are stacked.

Yield. Amount of grain harvested per unit of land area. In this handbook, all yields are expressed either in kilograms per hectare (kg/ha) or in metric tons per hectare (t/ha).

Annotated Bibliography

Listed below are many of the publications on which much of the information in this book is based. The bibliography bears directly on the subject of rice in the tropics and can provide an expanded and more detailed view of the topics covered. The literature citations are grouped by chapters. If a citation is listed under more than one chapter heading, the annotation is given only under the first chapter in which it appears.

Chapter 1: The Importance of Rice as a World Crop, and Its Principal Characteristics

Abelson, Philip H., ed. 1975. *Food: Politics, Economics, Nutrition and Research.* Washington, D.C.: American Association for the Advancement of Science. 202 p.

A compendium of articles on food production, most of which appeared earlier in a special edition of *Science.* The papers included are written by prominent authorities and constitute good background reading for those concerned about the world food problem.

American Society of Agronomy. 1975. *All-Out Food Production: Strategy and Resource Implications.* Special Publication No. 23. Madison, Wisconsin. 67 p.

Selected papers presented during the annual meeting of the American Society of Agronomy in 1974. Useful background reading on the world food problem.

Association of Japanese Agricultural Scientific Societies. 1975. *Rice in Asia.* Tokyo: University of Tokyo Press. 600 p.

Between 1966 and 1972 the Association of Japanese Agricultural

Scientific Societies held a series of symposia on "Rice in the World." This book is a collection of some of the papers that pertained to rice in Asia, particularly in South and Southeast Asia. The various chapters were contributed by Japanese scientists who had lived or traveled widely in Asia. An excellent assemblage of knowledge about rice in Asia up to 1972. Unfortunately, by the time the English edition was available, in late 1975, much of the information was out of date.

Brown, Lester R. 1974. *In the Human Interest.* New York: W. W. Norton and Company. 190 p.

The author suggests a strategy for stabilizing the world population, aiming at a maximum figure of 6000 million rather than the 10,000 million to 15,000 million forecast by other authorities.

Brown, Lester R., with Erik P. Eckholm. 1974. *By Bread Alone.* New York: Praeger Publishers. 272 p.

An analysis of the world's food resources in relation to population growth, with certain imperatives for adequate future food supplies.

Daly, Hermon E., ed. 1973. *Toward a Steady-State Economy.* San Francisco: W. H. Freeman Co. 332 p.

A collection of papers by various authorities, pertaining to the principle that the world economy eventually must reach some kind of equilibrium. The introduction by Professor Daly of Louisiana State University is especially useful to those concerned about the longtime solution to the problems of continued expansion in a finite world.

Food and Agriculture Organization. 1970. *Provisional Indicative World Plan for Agricultural Development.* 2 vols. Rome. 672 p.

Contains a synthesis and analysis of factors relevant to world, regional, and national development. Outlines the present state of agriculture in the less developed countries, predicts future food requirements and general needs for agricultural development, and makes suggestions for meeting those requirements.

Food and Agriculture Organization. 1971. *Food Balance Sheets. Average of 1964-66 Period.* Rome. 766 p.

Contains tables for most countries of the world, showing production and consumption of all major food products, processed and unprocessed. Includes the per capita intake of calories, proteins and fats from the various food sources.

Food and Agriculture Organization. 1976. *Production Yearbook.* Vol. 30. Rome. 296 p.

This yearbook appears about a year after the data are received by the FAO from the various countries. It contains the most complete assemblage of facts on area, yield, and production of about 100 crops on a country by country basis.

Grist, D. H. 1975. *Rice.* 5th ed. London: Longman Group Ltd. 601 p.

The fifth edition of the well-known book first published in 1953. A source of basic information on the rice plant, its history, culture and world importance. Unfortunately, its account of recent advances in rice research is incomplete.

Hopper, W. David. 1976. "The Development of Agriculture in Developing Countries." *Scientific American* 235, no. 3: 196-204.

An excellent analysis of the requirements for additional technology and for capital from the more developed countries to speed up agricultural development in the poorer nations.

Houston, D. F., and Kohler, G. O. 1970. *Nutritional Properties of Rice.* Washington, D.C.: National Academy of Sciences. 65 p.

A concise report on the vitamin, mineral, protein, and carbohydrate content rice and of its value as human food.

Houston, D. F., ed. 1972. *Rice Chemistry and Technology.* St. Paul, Minnesota: American Association of Agricultural Chemists. 517 p.

The most complete publication available on the properties, uses, and processing of the rice grain and its by-products. The contributors of the various chapters are among the foremost authorities in their fields of specialty.

International Rice Research Institute. 1977. *Constraints to High Yields on Asian Farms: An Interim Report.* Los Baños, Philippines. 235 p.

Information, presented at a workshop held in Yogyakarta, Indonesia, in 1976, on a cooperative study to determine why yields on farmers' fields in Asia are so far below those obtained in supervised experiments on the same fields. The studies, conducted in Thailand, Indonesia, and the Philippines from 1974 to 1976, are continuing and provide new insights into the constraints to high yields in the humid Asian tropics.

Mayer, Jean. 1976. "The Dimensions of Human Hunger." *Scientific American* 253, no. 3: 40-49.

A comprehensive description of the problem of hunger around the world, identifying the regions where malnutrition is the most severe.

Meadows, Donella H.; Meadows, Dennis L.; Randers, Jørgen; and Behrens, William W., III. 1972. *The Limits of Growth. A Report for The Club of Rome's Project on the Predicament of Mankind.* New York: Universe Books. 205 p.

An examination of the five basic factors that determine, and ultimately limit, growth on this planet: population, agricultural production, natural resources, industrial production, and pollution. Stimulating reading for those interested in life on this planet in the next century or two.

Palacpac, Adelita C. 1978. *World Rice Statistics.* Mimeographed. Los Baños, Philippines: International Rice Research Institute. 155 p.

Data on rice production, consumption, trade, prices, and fertilizer use, as well as other important statistics. Issued yearly.

Poleman, Thomas T., and Freebairn, Donald K., eds. 1973. *Food, Population and Employment. The Impact of the Green Revolution.* New York: Praeger Publishers. 272 p.

Papers presented at a workshop held at Cornell University in 1971 to explore the social, political, and economic consequences of the Green Revolution. The volume provides interesting and valuable reading because of the varying views of authorities from the several disciplines involved.

Scrimshaw, N. S., and Behar, M., eds. 1976. *Nutrition and Agricultural Development. Significance and Potential for the Tropics.* New York: Plenum Press. 500 p.

Papers presented and the ensuing discussion at a symposium on "Nutrition and Agricultural and Economic Development in the Tropics" held in Guatemala City in 1974. Contains papers by some of the world's most eminent authorities on human nutrition.

U.S. Department of Agriculture, Economic Research Service. 1975. *The World Food Situation and Prospects to 1985.* Foreign Agricultural Economic Report No. 98. Washington, D.C. 90 p.

An excellent analysis of the world food situation with special emphasis on cereal grain production and estimates of supply and demand during the next decade.

World Food Conference, Iowa State University. 1977. *Proceedings 1976*. Ames, Iowa: Iowa State University Press. 685 p.
    Contains 40 papers presented at the World Food Conference of 1976 at Ames, Iowa, and the final reports of six workshops that were held during the course of the 5-day conference. Although the subject matter covered varies greatly from paper to paper, there is much valuable information in the proceedings for those involved in agricultural development.

Wortman, Sterling. 1976. "Food and Agriculture." *Scientific American* 235, no. 3: 30-39.
    Treats the broad issues of the world food problem and is an introduction to a series of articles in this special issue of the *Scientific American* devoted to food and agriculture.

Chapter 2: The Modern Rice Plant and the New Technology: Greater Potentials for Rice Production in the Tropics

Chakkaphak, Chak. 1975. "Summary Report on Agricultural Mechanization and Development in Indigenous Farm Machinery Production in Thailand." *Agricultural Mechanization in Asia* 6, no. 2: 99-102.

Dalrymple, Dana G. 1976. *Development and Spread of High-Yielding Varieties of Wheat and Rice*. Foreign Agricultural Economic Report No. 95. Washington, D.C.: Economic Research Service, U.S. Department of Agriculture. 120 p.
    A compilation of the statistics and general information on the worldwide spread of the high-yielding wheat and rice varieties. Dalrymple has been updating these reports periodically, and undoubtedly others will be published in future.

International Rice Research Institute. 1963-77. *Annual Report*. 15 vols. Los Baños, Philippines.
    IRRI's annual reports give substantial accounts of its research activities and international programs. Usually there is a lag of little more than a year between the time the research is done and the work is published. However, IRRI puts out a smaller volume called *Highlights of Research* that appears about 6 months after the close of the year covered. This is a useful publication for busy administrators and gives a concise summary of the more important findings.

International Rice Research Institute. 1965. *The Mineral Nutrition of the Rice Plant*. Baltimore: Johns Hopkins Press. 494 p.

The proceedings of an IRRI symposium held in the Philippines in 1964. Contains the most complete information on the subject available at that time.

International Rice Research Institute. 1965. *The Rice Blast Disease*. Baltimore: Johns Hopkins Press. 507 p.

The proceedings of a symposium held at IRRI in 1963, to which the world's experts on the rice blast disease were invited. Still a valuable reference on the nature of the disease and methods of control.

International Rice Research Institute. 1967. *The Major Insect Pests of the Rice Plant*. Baltimore: Johns Hopkins Press. 729 p.

The proceedings of an IRRI symposium held in 1964. The most complete description, in one volume, of rice insects and their control that is available anywhere.

International Rice Research Institute. 1969. *Insect Pests of Rice*. Los Baños, Philippines. 78 p.

A booklet that concisely describes the major insect pests of rice.

International Rice Research Institute. 1969. *The Virus Diseases of the Rice Plant*. Baltimore: Johns Hopkins Press. 354 p.

With the exception of ragged stunt, which was not discovered until 1976, this book describes all the important virus diseases of rice in Asia. It includes the papers presented at a symposium on virus diseases of the rice plant held at IRRI in 1967.

International Rice Research Institute. 1972. *Rice Breeding*. Los Baños, Philippines. 738 p.

The proceedings of a symposium on rice breeding held at IRRI in 1971. The most comprehensive treatment of advances in rice breeding in the major rice-producing countries.

International Rice Research Institute. 1972. *Rice, Science and Man*. Los Baños, Philippines. 163 p.

The papers presented at the convocation celebrating the 10th anniversary of IRRI. The articles not only cover the progress made by the institute during its first decade but include subjects of wider scope, such as the role of international agricultural research institutes, the new rice technology and rural life, the outlook for world trade and rice production, and the economic consequences of the Green Revolution.

International Rice Research Institute. 1975. *Changes in Rice Farming in Selected Areas of Asia.* Los Baños, Philippines. 377 p.
The results of a study, initiated by IRRI in 1971, of the changes occurring in South and Southeast Asia as a result of the Green Revolution. The information was gathered from 36 rice-growing villages in 14 study areas in six countries.

International Rice Research Institute. 1975. *Major Research in Upland Rice.* Los Baños, Philippines. 255 p.
A summary of knowledge about upland rice and its culture as of 1973. The chapters were written separately by IRRI scientists and included the results not only of their research but that of others around the world. Provides useful information for improving the productivity of rice grown on small farms in rainfed areas.

International Rice Research Institute. In press. *Brown Planthopper: Threat to Rice Production in Asia.* Los Baños, Philippines.
This is a report of a symposium held at IRRI in 1977. It provides up-to-date information on all aspects of the brown planthopper as a pest of rice, including the problems of breeding resistant rice varieties.

International Rice Research Institute. 1977. *Constraints to High Yields on Asian Farms: An Interim Report.* Los Baños, Philippines. 235 p.
(See listing under chapter 1 for annotation.)

International Seminar on Deep-water Rice. 1975. *Proceedings.* Dacca: Bangladesh Rice Research Institute. 225 p.
The proceedings of the first attempt in recent years to bring together the best authorities on deep-water rice. The papers are informative; it was from this session that enthusiasm was generated for a cooperative research program on deep-water and floating rice.

Ling, K. C. 1972. *Rice Virus Diseases.* Los Baños, Philippines: International Rice Research Institute. 134 p.
An excellent description of the virus diseases of the rice plant.

Moseman, Albert H. 1971. *National Agricultural Research Systems in*

Asia. New York: Agricultural Development Council. 271 p.

Proceedings of a seminar sponsored by the Agricultural Development Council and held in New Delhi in 1970. A good account of the status of agricultural research in Asia in 1970 as reported by representatives from each country.

Ou, S. H. 1972. *Rice Diseases.* Kew, Surrey, England: Commonwealth Mycological Institute. 368 p.

The most complete treatment of the diseases of the rice plant that has appeared in the past 25 years. A useful reference for disease identification and control.

Ou, S. H. 1973. *A Handbook of Rice Diseases in the Tropics.* Los Baños, Philippines: International Rice Research Institute. 58 p.

A valuable practical guide to tropical rice diseases, with 21 color plates.

Palmer, Ingrid. 1975. *The New Rice in the Philippines.* Report Number 75.2. Geneva: United Nations Research Institute for Social Development. 199 p.

A socioeconomic study of the impact of modern rice varieties and technology on the rice industry in the Philippines.

Palmer, Ingrid. 1976. *The New Rice in Asia: Conclusions from Four Country Studies.* Report Number 76.5. Geneva: United Nations Research Institute for Social Development. 146 p.

An excellent summary of the four-country project. Highly recommended for those who cannot take the time to consult the individual country reports on the impact of the high-yielding varieties on the economy and general well-being of the people in India, Indonesia, the Philippines, and Sri Lanka.

Palmer, Ingrid. 1977. *The New Rice in Indonesia.* Report Number 77.1. Geneva: United Nations Research Institute for Social Development. 198 p.

A separate report for Indonesia of the four-country study made by the U.N. Research Institute for Social Development with UNDP financial support. An excellent analysis of the impact of the modern varieties on rice production in Indonesia. Includes an economic analysis of marketing and storage problems.

Philippine Council for Agriculture and Resources Research. 1977.

The Philippines Recommends for Rice 1977. Los Baños, Philippines. 186 p.
Contains the latest information on rice culture, including new varieties, cultural practices, the control of insects and diseases, and postharvest technology. Although designed for use in the Philippines, much of the information is applicable throughout the humid tropics. Updated frequently.

University of the Philippines, College of Agriculture (in cooperation with IRRI). 1970. *Rice Production Manual.* Los Baños, Philippines. 382 p.
Although now somewhat replaced by the *Training Manual for Rice Production* prepared by Xuan and Ross in 1976, this 1970 rice production manual contains much helpful information on rice culture in the tropics.

Xuan, Vo-Tong, and Ross, Vernon E. 1976. *Training Manual for Rice Production.* Los Baños, Philippines: International Rice Research Institute. 140 p.
A complete training manual for any practical course in rice production. Especially designed for training extension personnel and farm managers. Each lesson lists the materials and teaching aids needed and gives full instructions for conducting the exercises. The manual covers all operations from land preparation to drying and storage. For use in the less developed countries where labor is relatively abundant and farms are small.

Chapter 3: Problems of Postharvest Technology

Araullo, E. V.; de Padua, D. B.; and Graham, Michael, eds. 1976. *Rice Postharvest Technology.* Ottawa: International Development Research Centre. 394 p.
A compendium of the material used at a postharvest technology training course held in Los Baños, Philippines, in 1973. A valuable source of basic information on all postharvest operations from cleaning, drying, and storage to parboiling and milling. Contains designs of equipment and states the options available to countries wishing to improve and expand their facilities.

Houston, D. F., ed. 1972. *Rice Chemistry and Technology.* St. Paul, Minnesota: American Association of Agricultural Chemists. 517 p.
(See listing under chapter 1 for annotation.)

Timmer, C. Peter. 1972. "Employment Aspects of Investment in Rice Marketing in Indonesia." *Stanford Food Research Institute Studies* 11, no. 1: 59-88.

An analysis of different milling techniques with special reference to labor-surplus Indonesia.

Chapter 4: Rice Marketing

Abbott, J. C.; Barter, P. G. H.; Kelly, R. W.; and Spinks, G. R. 1972. *Rice Marketing.* FAO Marketing Guide No. 6. Rome: Food and Agriculture Organization. 189 p.

A thorough discussion of the problems of rice marketing around the world, with many specific examples of the widely varying marketing procedures in various countries.

Asian Development Bank. 1977. *Asian Agricultural Survey 1976.* Manila. 490 p.

A well-written and highly informative report of the state of agricultural development in South and Southeast Asia as of 1976. Analyzes the problems and gives sound advice on the more promising approaches to their solution. Good reading for all Asian agricultural administrators, planners, and developers.

Efferson, J. Norman. 1972. "Outlook for World Rice Production and Trade." In *Rice, Science and Man,* pp. 127-42. Los Baños, Philippines: International Rice Research Institute.

International Rice Research Institute. 1971. *Viewpoints on Rice Policy in Asia.* Los Baños, Philippines. 275 p.

Mimeographed papers presented at a rice policy conference held at IRRI in 1971. A summary of and general perspective on the conference, prepared by Vernon W. Ruttan, is especially useful to those who do not wish to read all the papers.

Chapter 5: Some Successful Rice Production Programs

Drilon, J. D., Jr. 1975. "An Overview of Masagana 99." *Modern Agriculture and Industry* 3, no. 12: 14-16, 87-89.

A description of the Masagana 99 program and of its achievements during the first 2 years of operation.

Food and Agriculture Organization. 1976. *Production Yearbook.*

Vol. 30. Rome. 296 p.
(See listing under chapter 1 for annotation.)

Korea Development Institute. 1975. *Korea's Economy, Past and Present*. Seoul. 367 p.
A good description of South Korea's recent development, including that of agriculture.

Korea, Office of Rural Development. 1976. *Rural Development Program in Korea—1976*. Suweon, Korea.
An excellent description, with many illustrations, of how South Korea became self-sufficient in rice production.

Li, K. T. 1977. "Strategy for Rice Production in Taiwan." In *Proceedings, Food Crisis Workshop*, pp. 141-170. Manila: Ramon Magsaysay Award Foundation.
An excellent account of Taiwan's policies for maintaining self-sufficiency in rice production.

Mellor, John W. 1976. "The Agriculture of India." *Scientific American* 235, no. 3: 154-163.
A good analysis of India's food production problems. Done in a more optimistic vein than are similar articles by several other authors.

Philippines, Department of Agriculture, and National Food and Agriculture Council. 1976. *Masagana 99. A Program of Survival*. Quezon City, Philippines. 33 p.
A brief description of the basic elements in the Masagana 99 program to attain rice sufficiency in the Philippines.

Scobie, Grant M., and Posada, Rafael T. 1977. *The Impact of High-Yielding Rice Varieties in Latin America, with Special Emphasis on Colombia*. Cali, Colombia: Centro Internacional de Agricultura Tropical. 163 p.
A thorough analysis of the rice industry in Colombia, documenting the impact on average yields of modern rice varieties on irrigated land.

Shen, T. H. 1976. *Taiwan's Family Farm During Transitional Economic Growth*. Ithaca, New York: Program in International Agriculture, Cornell University. 14 p.
Describes briefly the changes taking place in Taiwan as

industrialization increases. Gives data on mechanization, on joint farming, operations, and on specialized farming areas in the region.

Shin, Dong Wang, and Shim, Yong Kun. 1975. *The Effectiveness of the Tongil Rice Diffusion in Korea.* Suweon, Korea: Korea, Office of Rural Development. 52 p.

An excellent discussion of the way in which South Korea added a ton per hectare to its average rice yield through breeding and distributing new rice varieties and by introducing further improvements in rice cultivation methods.

Tanco, Arturo R., Jr., and Feuer, Reeshon. 1976. "Philippine Rice Self-Sufficiency through Masagana 99. An Example of the Successful Transfer of Technology to Small-Scale Rice Farmers." *International Rice Commission Newsletter* 25, No. 1/2: 29-30.

Briefly describes the achievements of the Masagana program 3 years after its beginning. The authors list the essential ingredients of the program, which have broad application especially in Southeast Asia.

Yang, Y. K. 1977. *Farmers' Organizations in Taiwan.* Taipei, Taiwan: Joint Commission on Rural Reconstruction. 26 p.

A clear description of the farmers' associations and irrigation associations of Taiwan.

Chapter 6: Promising Rice Research

Brown, A. W. A.; Byerly, T. C.; Gibbs, M.; and San Pietro, A., eds. 1976. *Crop Productivity . . . Research Imperatives.* East Lansing, Michigan: Michigan State University. 399 p.

The conclusions of six working groups and the thoughts of a number of prominent authorities who participated in an international conference on crop productivity, held at Harbor Springs, Michigan (U.S.) in 1975. The principal objective of the conference was to identify both short-term and long-term priorities for research to increase crop productivity in the less developed nations. Valuable reading for research administrators seeking the most promising and significant research areas that will have an impact on national rice yields.

Council for Asian Manpower Studies. 1975. "Multiple Cropping in Asian Development," *Philippine Economic Journal* 14, No. 1/2: 1-322.

These first two numbers of the *Philippine Economic Journal* for 1975 are devoted to the papers given at a conference on multiple cropping held in Taipei in late 1973 and sponsored by the Council for Asian Manpower Studies. Because the speakers, who were from Southeast Asian countries and Taiwan, dealt largely with multiple cropping systems that included rice, this material will be of interest to administrators desiring to increase food production either by growing more than one crop of rice a year or by adding other crops in the rotation. Gives an especially good review of multiple cropping in Taiwan, where it has been so successful.

Cummings, Ralph W., Jr. 1976. *Food Crops in the Low-Income Countries: The State of Present and Expected Agricultural Research and Technology.* Working Papers. New York: Rockefeller Foundation. 103 p.
Analyzes the present state of research on and knowledge of the world's major food crops. The work of the international agricultural research institutes with these crops is discussed and those that need additional emphasis are identified.

Dalrymple, Dana G. 1971. *Survey of Multiple Cropping in Less-Developed Nations.* Foreign Economic Development Report 12. Washington, D.C.: Foreign Economic Development Service, U.S. Department of Agriculture. 108 p.
An excellent review of the history of the development of multiple cropping and of its present status in some 25 countries. Concludes with a discussion of future prospects and policy issues.

Deep-water Rice Workshop. 1977. *Proceedings.* Los Baños, Philippines: International Rice Research Institute. 239 p.
Papers presenting current knowledge about deep-water rice. Includes information useful to agricultural administrators and scientists in countries where deep-water rice is an important crop.

International Rice Research Institute. 1963-77. *Annual Report.* 15 vols. Los Baños, Philippines.
(See listing under chapter 2 for annotation.)

International Rice Research Institute. 1972. *Rice, Science and Man.* Los Baños, Philippines. 163 p.
(See listing under chapter 2 for annotation.)

International Rice Research Institute. 1975. *Major Research in Upland Rice.* Los Báñôs, Philippines. 255 p.
(See listing under chapter 2 for annotation.)

International Rice Research Institute. In press. *Soils and Rice.* Los Baños, Philippines.
Proceedings of a conference on soils and rice held in the Philippines in 1977. An excellent report of the latest scientific advances regarding the characteristics and management of soils on which rice is grown.

International Seminar on Deep-Water Rice. 1975. *Proceedings.* Dacca: Bangladesh Rice Research Institute. 225 p.
(See listing under chapter 2 for annotation.)

Matsushima, Seizo. 1976. *High-Yielding Rice Cultivation.* Tokyo: University of Tokyo Press. 367 p.
An account of the author's views on the management of the rice plant for maximum yield. The latest of several books he has published on the same general subject. His theories are backed by practical experience in obtaining yields in excess of 10 t/ha of paddy.

Moseman, Albert H. 1971. *National Agricultural Research Systems in Asia.* New York: Agricultural Development Council. 271 p.
(See listing under chapter 2 for annotation.)

Ranit, Luis C., with J. D. Drilon, Jr. 1977. "Lorenzo P. Jose Rice Farm: A "Computerized" Japanese Type Rice Farming Enterprise." In *Proceedings, Food Crisis Workshop,* pp. 261-301. Manila: Ramon Magsaysay Award Foundation.
A fascinating account of a Filipino farmer who devised a system of continuous rice cultivation on 1.5 hectares of land. The average yield was approximately 26 t/ha annually, accomplished with family labor only.

Chapter 7: Elements of a Successful Accelerated Rice Production Program
Asian Development Bank. 1977. *Asian Agricultural Survey 1976.* Manila. 490 p.
(See listing under chapter 4 for annotation.)

Benor, Daniel, and Harrison, James Q. 1977. *Agricultural Extension. The Training and Visit System.* Washington, D.C.: World Bank. 55 p.

A clearly written, concise bulletin that outlines a system of agri-extension particularly adapted to the poorer countries attempting to move their agriculture from the traditional to the modern in a practical way.

Boyce, James K., and Evenson, Robert E. 1975. *National and International Agricultural Research and Extension Programs.* New York: Agricultural Development Council. 229 p.

Contains the results of a thorough study of the investments in agricultural research and extension in recent years and of the contribution those efforts have made toward agricultural development.

Chandler, Robert F., Jr. 1977. "Some Thoughts on Accelerating Food Production in the Less Developed Countries." In *Proceedings, Food Crisis Workshop,* p. 57-66. Manila: Ramon Magsaysay Award Foundation.

A discussion of the more important factors determining national rice yields in Asia.

Cheany, Robert L., and Jennings, Peter R. 1975. *Field Problems of Rice in Latin America.* Cali, Colombia: Centro Internacional de Agricultura Tropical. 91 p.

A manual designed to assist farmers and field technicians in identifying the more common insects and diseases and the plant symptoms of adverse soil conditions in Latin America.

International Rice Research Institute. 1975. *Water Management in Philippine Irrigation Systems: Research and Operations.* Los Baños, Philippines. 270 p.

The proceedings of a water management workshop held in 1972 and sponsored jointly by the IRRI and the College of Agriculture of the University of the Philippines. Among other points, it identifies the weaknesses in present irrigation systems, particularly in their management. Although it contains only Philippine data, the information has much broader application, especially in South and Southeast Asia.

Jennings, Peter R. 1976. "The Amplification of Agricultural Production." *Scientific American* 235, no. 3: 180-195.

The role that improved varieties of rice and wheat can play in increasing future food supplies. An excellent article.

Mueller, K. E. 1970. *Field Problems of Tropical Rice.* Los Baños: International Rice Research Institute.

A pocket-size manual containing color photographs and brief descriptions of common rice insects and of symptoms of diseases and soil problems.

Okita, Saburo, and Takase, Kunio. 1977. "Doubling Rice Production in Asia." In *Proceedings, Food Crisis Workshop,* pp. 187-217. Manila: Ramon Magsaysay Award Foundation.

The theme of this article is that until Asia invests large sums of money in rice irrigation projects yields will remain low, but that by irrigating the suitable areas average rice production can be doubled in the next two decades.

Chapter 8: A National Rice Program—Putting the Ingredients Together

Adams, Dale W., and Coward, E. Walter, Jr. 1972. *Small-Farmer Development Strategies: A Seminar Report.* New York, N.Y.: Agricultural Development Council. 33 p.

A summary of the papers presented and the principal ideas expressed at a seminar held at Ohio State University in 1971 that was designed to explore the most appropriate means of reaching the small farmer in agricultural development programs.

All-India Coordinated Rice Improvement Project. 1974. *India's Rice Revolution. A Beginning.* Hyderabad, India. 72 p.

The achievements of the All India Coordinated Rice Improvement Project during its first 8 years of operation (1965 to 1973). A good example of what can be done to coordinate local research programs into a successful national effort.

Brown, Lester, R. 1970. *Seeds of Change.* New York: Praeger Publishers. 205 p.

A thought-provoking book, showing the impact of the modern varieties of rice and wheat but also identifying the problems that lie ahead in maintaining the increased tempo that the new varieties initiated.

Castillo, Gelia T. 1975. *All in a Grain of Rice*. Los Baños, Philippines: Southeast Asian Regional Center for Graduate Study and Research in Agriculture. 410 p.

Written by a well-known rural sociologist of the University of the Philippines at Los Baños, this work gives a highly objective analysis of the impact of the modern rice varieties and the new technology on the Filipino farmer. In addition, it presents and analyzes the views of others who have written significant papers on the Green Revolution.

Chambers, Robert. 1974. *Managing Rural Development. Ideas and Experience from East Africa*. Uppsala: Scandinavian Institute of African Studies. 216 p.

Reflects the author's experience in Africa while he was engaged in rural development activities, especially in Kenya. Clearly written, it stresses the programing and management needed for properly transferring information obtained by the research scientists to the extension staff, with final application at the farm level.

Mosher, A. T. 1966. *Getting Agriculture Moving*. New York: Praeger Publishers. 191 p.

This book, which has been translated into a number of languages, contains a discussion of the essentials for developing and modernizing agriculture in the less developed countries. It was written as a text and reference work for in-service training programs. However, it is useful reading for administrators and planners involved in agricultural improvement.

Mosher, A. T. 1969. *Creating a Progressive Rural Structure*. New York: Agricultural Development Council. 172 p.

An excellent treatise on the steps to be taken by agricultural and development officers in organizing agricultural improvement programs. A straightforward and practical book that should be required reading for everyone involved in boosting the productivity and economy of rural areas in the less developed countries.

Mosher, A. T. 1971. *To Create a Modern Agriculture*. New York: Agricultural Development Council. 162 p.

This book grew out of a series of lectures given by the author in India in 1971, which became the basis of seminars in other Asian countries. The book synthesizes the lecture material with a reflection of audience comment at both lectures and seminars.

Mosher describes agriculture as an industry and outlines the steps of organizing and planning a program of agricultural development.

This volume should be read in conjunction with the three other works of the author that are listed here.

Mosher, A. T. 1975. *Serving Agriculture as an Administrator*. New York: Agricultural Development Council. 64 p.

Sets forth the principles of agricultural administration, pointing out the crucial role that agricultural officers play in making any development program successful.

Poleman, Thomas T., and Freebairn, Donald K., eds. 1973. *Food, Population and Employment. The Impact of the Green Revolution*. New York: Praeger Publishers. 272 p.

(See listing under chapter 1 for annotation.)

Rao, V. K. R. V. 1974. *Growth with Justice in Asian Agriculture. An Exercise in Policy Formation*. Report Number 74.2. Geneva: United Nations Research Institute for Social Development. 95 p.

An analysis of agricultural policy formation in Asia based on interviews with top officials and scientists in the FAO (Rome) and in the governments of Japan, Malaysia, the Philippines, Thailand, India, and Bangladesh. Useful reading for agricultural officers concerned with the problems of improving the lot of the small-scale farmer in a subsistence-farming economy.

Appendix

Brady, Nyle C. 1976. "Rice Research and Training in International Agricultural Research Centers." *International Rice Commission Newsletter* 25, No. 1/2: 6-28.

A concise but complete description of the rice research and training programs at the IRRI, IITA, and CIAT. It is well worth reading by the busy administrator.

International Agricultural Development Service. 1977. *First Report/1976*. New York. 81 p.

Describes the policies, objectives, and achievements of IADS during its first year of activity.

International Agricultural Development Service. 1978. *Agricultural Assistance Sources*. New York. 149 p.

Contains brief descriptions of the activities and interests of 20 organizations that offer financial and technical assistance to developing countries. It is an up-to-date and highly useful source of information. Unfortunately, it does not contain information about Japanese organizations, which would be of special importance to Asian rice-growing nations.

International Bank for Reconstruction and Development. 1975. *Rural Development: A Sector Policy Paper.* Washington, D.C.: World Bank. 89 p.
Discusses the nature and extent of rural poverty, and the policies and programs that can be set up to promote rural development in the poor countries. It also describes the World Bank program for promoting rural development, particularly in countries and regions where small-scale, subsistence farming has been the pattern in the past.

World Bank. 1976. *Questions and Answers.* Washington, D.C.: World Bank. 71 p.
A most useful publication for administrators in nations considering approaching the World Bank for a development loan. It explains how to start negotiations and describes the kinds of projects in which the bank has a major interest.

Index

Afghanistan, 2

Africa, 2, 12, 14, 20, 24, 93, 96, 138, 142, 149, 150, 181; area, yield, and production of rice in, 7; potential for rice production in, 96

African Development Bank, agricultural programs of, 215

Agricultural Development Council, programs of, 212

Amino acids, content of: in rice grain, 10

Area planted to rice, by continents and countries, 1-9

Armyworm, 46

Asian countries: area, yield, and production of rice in, 2-7

Asian Development Bank, 140, 160, 168, 183; agricultural programs of, 214

Australia, 7, 13; foreign assistance agencies of, 224

Azolla sp. (water fern), 42, 156, 157

Bacterial blight, 52

Bacterial streak, 52

Bali, 14, 23

Bangladesh, 3, 16, 20, 62, 150, 159, 165, 166

Basmati rice, 16, 140

Belgium, foreign assistance agency of, 223

Bran, uses of, 85

Brazil, 7, 149

Broken rice, uses of, 85-86, 98

Brown planthoppers, 46, 47, 49, 52

Brown rice, 10, 11

Brown spot disease, 51

Buffer stocks, 92

Burma, 3, 22, 93, 159, 201; Agricultural Marketing Board of, 93

Cambodia. *See* Kampuchea

Canadian International Development Agency (CIDA), agricultural programs of, 221-222

Carbofuran, 49, 157

Carbohydrate in rice grain, 10-11

Central America, 102, 138. *See also* Latin America

Central Rice Research Institute (India), 33

Centro Internacional de Agricultura Tropical (CIAT). *See* International Center for Tropical Agriculture

Chemical composition, of rice grain, 9

Chilo suppressalis. See Rice stem borers

China, 2, 9, 13, 16, 18, 22, 23, 25, 27, 155, 159, 160, 174, 188, 190, 199; communes serving as farmers' organizations in, 199; future rice production in, 23

China, Republic of. *See* Taiwan

CIAT. *See* International Center for

Tropical Agriculture

CIDA. *See* Canadian International Development Agency

Cleaning and drying operations, economics of, 72-74

Cleaning rice. *See* Rice cleaners

Climatic conditions, suitability of: for rice, 170

Cnaphalocrosis medinalis. *See* Rice leaf folder

Cold tolerance of rice, 18

Colombia, 7, 24, 102, 142, 175; agricultural credit in, 138; the Caja Agraria in, 138; the Fondo Financiero Agrario in, 138; inputs used on irrigated land in, 136-137; investment in irrigation by, 134; irrigated rice in, 132; National Federation of Rice Growers in, 131, 136, 138; rice breeding program in, 130-132; rice consumption in, 137; rice production program of, 129-139; rice yields in, 129-130; support price for rice in, 137, 138; transferability of experience of, 138-139; yields of irrigated and upland rice in, 132-134

Colombian Department of Agriculture (ICA), 131

Colombian National Federation of Rice Growers. *See* Colombia, National Federation of Rice Growers in

Commercial companies, assistance from, 224

Cooperatives, 68, 92, 200

Corticum saskii. See Sheath blight

Credit, rural, 182-184; interest rates for, 187-188; supervision of, 183

Cropping systems research, 160-162

Cyperus rotundus. See Nutsedge

Danish International Development Agency, 223

Deep-water rice, definition and culture of, 20

Demand for rice, 94, 181

Demonstrations, on-farm, 172, 173, 191, 192

Denitrification, 41

Diseases, of rice, 51-53, 117

Drying rice. *See* Rice dryers

Economic and social studies of new rice technology, 61-64

Economy of scale, 195

Efferson, J. Norman, 91 n

Egypt, 13

Employment: off-farm, 186-188; in public works, 187-188; in rural manufacturing industries, 188

Europe, 7, 79, 80, 87, 168

European Development Fund (EDF), agricultural programs of, 217

Export markets: developing new, 99; internal improvements needed for development of, 99-100; potentials of, 96-97

Extension field staff: importance of farm visits by, 179; maintaining morale of, 181; side roles of, 180; training of, 177-178

Extension service: field days conducted by, 179; need for improvement of, 177; need for unified command in, 179-180. *See also* Extension field staff

FAO. *See* Food and Agricultural Organization

Farm trials, 172, 173

Farmers' associations, 92; in Taiwan, 109-110, 200

Farmers' organizations, 142, 198-200; in Taiwan, 109-112

Farming district: definition of, 191; organizing of, 191-193

Farming locality: definition of, 190; organizing of, 190-191

FEDEARROZ. *See* Colombia,

National Federation of Rice Growers in

Fertilizer, 3, 62-63, 173-174, 196-197; organic, 154-155; response of modern varieties to, 38-39

Field resistance. *See* Rice breeding, for horizontal resistance

Floating rice. *See* Deep-water rice

Flood control, 168

Food and Agricultural Organization of the United Nations (FAO), 3, 9, 21, 23, 33; Intergovernmental Group on Rice of, 87; programs of, 215-216

Ford Foundation, agricultural programs of, 211

Foreign aid, 93

France, 13; foreign assistance agency of, 223

Fungicides, 198

Gall midge, 46, 47

Germany, Federal Republic of: assistance agencies in, 220

Glutinous rice, 9-10, 15, 16

Grades and standards: importance of using, 87; for paddy, 73

Grassy stunt disease, 47, 52

Green leafhoppers, 46, 47, 49, 52

Green Revolution, 2, 62, 204

Handling and transporting rice, economics of, 74-75

Harvest and postharvest operations, losses during, 65, 66, 74, 75

Harvesting, timing of, 66

Helminthosporium oryzae. *See* Brown spot disease

Herbicides, 176, 197; butaclor, 53; thiobencarb, 53; 2,4-D, 53, 54; use of, on upland rice, 55

Hispa amigera. *See* Rice hispa

Hoja blanca disease, 52, 53, 130, 131

Hopperburn, 47

Hulls, uses of, 84-85

Husks. *See* Hulls

Hydrellia philippina. *See* Whorl maggot

IADS. *See* International Agricultural Development Service

IBRD. *See* International Bank for Reconstruction and Development

ICA. *See* Colombian Department of Agriculture

IDA. *See* International Development Association

IDRC. *See* International Development Research Centre

IITA. *See* International Institute of Tropical Agriculture

India, 2, 9, 16, 20, 33, 47, 51, 56, 60, 62, 79, 150, 159, 165, 166, 171; Food Corporation of, 93; increases in rice yields in, 140-141

Indica-japonica hybridization, 33

Indonesia, 2, 16, 20, 23, 52, 62, 159, 174; increases in rice yields in, 140; National Logistic Authority of, 93

Inputs: need for availability of, to farmers, 173; subsidization of, 185-186

Insect control: with insecticides, 48-49; by integrated pest control methods, 50; reducing cost of, 157-158; by using resistance varieties, 46-47

Insecticides, 175-176, 197; systemic, 49; use of, 48-49, 63

Insects attacking rice, 46; biological control of, 46; varietal resistance to, 46-47

Inter-American Development Bank, agricultural programs of, 214-215

International Agricultural Development Service (IADS), programs of, 212-213

International Bank for Reconstruction and Development (IBRD), 160; agricultural programs of, 213-214

International Center for Tropical Agriculture (CIAT), 131, 138, 139, 161, 162, 201; rice program at, 209-210

International Development Association (IDA), 213

International Development Research Centre (IDRC), agricultural programs of, 221-222

International Fertilizer Development Center, 41

International Institute of Tropical Agriculture (IITA), 150, 161, 201; rice program at, 210

International Rice Research Institute (IRRI), 14, 25, 31-32, 33, 34, 36, 38, 41-42, 45, 55, 57-59, 60, 61, 62, 117, 118, 124, 128, 131, 146, 148, 149, 150, 151, 152, 154, 161, 162, 201, 202; activities of, 207-209; International Rice Testing Program of, 208; training program of, 208

Iran, 2

IR8, 34-38, 47, 117, 131, 140, 148

IR8-288-3. *See* IR8

IRRI. *See* International Rice Research Institute

Irrigated lowland rice, definition and culture of, 18

Irrigation, 3; in Colombia, 132-134; importance of, 62, 196; in Philippines, 128, 129; in South Korea, 116; in Taiwan, 104, 110-111

Irrigation associations, 199; in Taiwan, 110-111

Irrigation systems: design and management of, 168-169; in Philippines, 129

Italy, 13

Ivory Coast, 7, 14, 141, 150

Japan, 2, 13, 18, 22, 23, 25, 31, 33, 51, 56, 95, 103, 104, 114, 116, 154, 159, 168, 184, 195, 196, 197, 198; foreign assistance agencies of, 222-223

Java, 14, 23

Kampuchea, 3, 93, 123, 159

Korea. *See* South Korea

Korea, Democratic People's Republic of. *See* North Korea

Korea, Republic of. *See* South Korea

Korea, South. *See* South Korea

"Kresek." *See* Bacterial blight

Land preparation: economics of, 58-59; equipment for, 57-59

Laos, 3, 7, 16, 93, 123, 159; level of rice yield in, 141

Latin America, 16, 24, 53, 93, 102, 132, 139, 142; potential for rice production in, 96; rice pricing policies in, 95

Leptocorisa acuta. See Rice bug

Loans to farmers, problems of collecting, 183

Lysine, 10

Machinery for small farms, 56-61. *See also* Power equipment

Madagascar, 14

Madura, 23

Malaysia, 2, 16, 33, 52, 62, 95, 159, 165; Padi and Rice Marketing Board of, 93

Marketing of rice: export, 96-100; local, 92-93, 100

Masagana 99 program, 124-128, 129; credit aspects of, 125-126; extension activities of, 126-127, 180; impact on rice yields of, 127; loan repayment rates in, 128; subsidization of fertilizer in, 127; support price for rice in, 127

Mechanization of rice industry: appropriateness of, 194-196

"Mentek." *See* Tungro disease
Methionine, 10
Mimosa invisa, 158
"Miracle rice," 204
Mosher, A. T., 190
Multiple cropping. *See* Cropping systems research
Mud balls, application of nitrogen in, 41

National rice programs, selection of land areas for, 193-194
Natural resources, analyzing of, 167-170
Nepal, 3, 159
Nephotettix nigropictus. See Green leafhoppers
Nephotettix virescens. See Green leafhoppers
The Netherlands, foreign assistance agencies of, 222
New Zealand, foreign assistance agency of, 224
Nigeria, 7, 16, 166
Nilaparvata lugens. See Brown planthoppers
Nitrogen, biological fixation of, 41-42, 153, 156-157
Nitrogen fertilizer: losses of, from soil, 40-41; from organic sources, 154-155; response of modern varieties to, 38-39
North Korea, 2
Norwegian Agency for International Development, 224
Nutritional value of rice, 9-12
Nutsedge, 55, 158
Nymphula depunctalis. See Rice caseworm

Pachdiplosis oryzae. See Gall midge
Pakistan, 2, 16, 22, 27, 93, 159; progress in raising rice yields in, 140
Parboiled rice, 11, 16, 79
Parboiling of rice, 79-80

"Penyakit merah." *See* Tungro disease
People's Republic of China. *See* China
Pest control, integrated methods of, 197
Pest populations, need for estimate of, 170-171
Philippine National Bank, 125
Philippines, 2, 3, 7, 9, 14, 16, 22, 38, 47, 53, 54, 55, 56, 57-59, 62, 102, 142, 154, 159, 161, 165, 174, 175, 183, 196, 201; Agricultural Credit Administration of, 125; Bureau of Agricultural Extension of, 124; expansion in rice irrigation in, 128; management of irrigation systems in, 129; National Food and Agricultural Council of, 124; National Grain Authority of, 93; rice production program of, 123-129; rice yields in, 123
Phosphorus, 39-40, 62, 106, 153, 174, 197
Plant type, 34-35
Planthopper. *See Sogatodes oryzicola; see also* Brown planthoppers
Policies from abroad, usefulness of, 194-200
Polygenic resistance. *See* Rice breeding, for horizontal resistance
Ponlai rice, 105
Population, estimates of future global, 24-25, 29
Population growth rate, need for controlling, 205
Portugal, 13
Potassium, 39-40, 63, 106, 174, 197
Power equipment, need for: in rice production, 176
Power tillers, 57-59
Price controls. *See* Price supports
Price ratio of nitrogen to paddy, 197

Price supports, 95, 184-185; in Colombia, 137, 138; determining appropriate policies for, 198; in Philippines, 127; in South Korea, 118-119; in Taiwan, 109
Production. *See* Rice production
Protein in rice grain, 10, 11
Pseudoletia unipuncta. *See* Army-worm
Pyricularia oryzae. *See* Rice blast disease

Quality control. *See* Grades and standards
Qualities of rice: consumer preferences for, 15-16; cooking and eating, 15-16; for export markets, 98

Ragged stunt disease, 47, 52, 53
Rainfed paddy, description and culture of, 18-19
Research. *See* Rice research, importance of
Rice blast disease, 51, 115
Rice bran. *See* Bran
Rice breeding, 33-35; for drought tolerance, 20, 149-150; for early maturity, 146; for fertilizer responsiveness, 148; gene pyramiding in, 147; for horizontal resistance, 147; multiline varieties development in, 147; for stable resistance to insects and diseases, 146-147; for tolerance to adverse soil conditions, 152-153; for tolerance to varying water depths, 150-152
Rice bug, 46
Rice by-products, 84-86
Rice caseworm, 46
Rice cleaners, selection of, 68-69
Rice consumption: in Colombia, 137; by Europeans, 9; in principal Asian countries, 9; in United States, 9
Rice distribution, increasing the

efficiency of, 86-87
Rice dryers, 71, 72
Rice hispa, 46
Rice leaf folder, 46
Rice milling, 81
Rice mills: capacity of, 83-84; economics of operation of, 83, 89; outturn of, 81; types of, 81-83
Rice plant: development of modern types of, 32-38; traditional tropical, 32, 34; unique characteristics of, 17-18
Rice prices, world, 97-98. *See also* Price supports
Rice production: by continents and countries, 1-9; estimates of future, 25-29; system for continuous, 162-163
Rice production programs, maintaining the pace of, 204-205
Rice production training: at the International Center for Tropical Agriculture (CIAT), 210; at the International Institute of Tropical Agriculture (IITA), 210; at the International Rice Research Institute (IRRI), 208
Rice research, importance of: in national programs, 171, 200-201
Rice self-sufficiency, as national goal, 94-95
Rice soils, chemical changes in: after flooding, 43-44
Rice starch, amylose and amylopectin in, 9, 15-16
Rice stem borers, 46, 47, 49
Rice types: bulu, 14; indica, 12-13; japonica, 13; *Oryza glaberrima*, 14-15
Rice varieties: ADT 27, 33; Bluebonnet 50, 130, 131, 132; BPI-76, 33; Chianan 8, 105; Chianung 242, 105; CICA 4, 131; CICA 6, 131; CICA 7, 131; CICA 9, 131, 132; Dee-geo-woo-gen, 34, 36; Gulfrose, 130; H-4, 33; H-5, 33;

I-geo-tse, 34; IR20, 38, 153; IR22,
131; IR26, 47; IR28, 153; IR29,
153; IR30, 152, 153; IR32, 152;
IR34, 153; IR36, 47, 146, 152;
IR38, 47; IR40, 47; IR42, 47, 149;
Kaohsiung, 53, 105; Malinja, 33;
Mashuri, 33; Milyang 21, 118;
Milyang 23, 118; Napal, 130;
Peta, 34, 38, 47; Taichung Native
1, 34, 36, 105, 117; Taichung 65,
105; Tainan 1, 105; Tainan 3,
105; Tainan 5, 105; Tapuripa,
130, 131; TKM6, 149; Tongil,
117; Yukara, 117, 118; Yushin,
118. *See also* IR8
Rice yields: in Colombia, 129-130;
by continents and countries, 1-9;
in Philippines, 123-124; reasons
for low, 159-160; in South Korea,
113-114; in Taiwan, 102-104. *See
also* Yield potentials
Roads, farm-to-market, 181
Rockefeller Foundation, agricultural programs of, 211

Saemaul Undong. See South
Korea, New Village Movement
Seed, 174-175
Seeding equipment, 59-60, 176
Senegal, 166
Seoul National University, College
of Agriculture of, 117
Sesamia inferens. See Rice stem
borers
Sheath blight, 51
Silica, 118
Silicon, 17, 40
Sogatella furcifera. See White-
backed planthopper
Sogatodes oryzicola, 52, 130
Soil conditions, need for survey of,
169-170
Soil nitrogen, 41
Soil pH, effect of flooding on, 17
Soil salinity, tolerance of rice to, 17
Solar radiation, effect on rice yield

of, 44-45
South America, 2, 7, 102, 138; area,
yield and production of rice in,
5, 7. *See also* Latin America
South Korea, 2, 13, 18, 23, 25, 56, 95,
102, 142, 150, 167, 175, 105, 106,
197, 198, 200, 204; comparison of
rural and urban incomes in, 121;
cooperative farming units in,
120; development of Tongil rice
in, 117-118; fertilizer use in, 115-
116; New Village Movement in,
119-121, 200; Office of Rural Development of, 116, 117, 120; price
incentives in, 118-119; rice cultural practices in, 118; rice irrigation in, 116; rice production program in, 113-122; rice yields in,
113-114, 122; varietal improvement in, 116-118
Soviet Union, 8, 13; area, yield and
production of rice in, 8; increases
of rice yields and production in,
141-142
Spain, 13
Sri Lanka, 3, 16, 19, 33, 89, 159;
Paddy Marketing Board of, 93
Storage facilities: bird-proofing of,
78; control of moisture in, 78;
fumigation of, 78; need for, for
buffer stocks, 92; planning of, 77;
temporary, 77; types of, 76
Storage, losses during, 75, 78
Stipe virus disease, 117
Surinam, 131
Swedish International Develop-
ment Authority, 224
Swiss Technical Cooperation, 224
Systems approach, use of: in post-
harvest operations, 89-90

Taiwan, 2, 13, 31, 34, 54, 56, 57, 95,
119, 121, 122, 136, 142, 167, 175,
190, 195, 196, 197, 198, 199, 200;
farmers' associations in, 109-110;
fertilizer use in, 103, 106; guaran-

teed minimum rice prices in, 109; irrigation associations in, 110-111; joint farming in, 111-112; land consolidation in, 111-112; land reform in, 108; off-farm employment opportunities in, 112-113; per capita income in, 112; pesticide use in, 106; rice-fertilizer barter system in, 107; rice irrigation in, 104; rice production program in, 102-113; subsidization of inputs in, 108; varietal improvement in, 104-106; yield of paddy in, 102-104

Thailand, 3, 20, 22, 23, 24, 27, 62, 87, 93, 95, 119, 150, 151, 159, 196, 201; development of rice machinery in, 56-57; Rice Traders Association of, 87

Threonine, 10

Threshing equipment, 60-61, 66

Timmer, C. Peter, 83

Tryporyza incertulas. See Rice stem borers

Tungro disease, 52

UNDP. *See* United Nations Development Programme

United Kingdom, Ministry of Overseas Development of, 219-220

United Nations, 24, 29

United Nations Development Programme (UNDP), 216-217

United Nations Research Institute for Social Development, 62

United States, 7, 9, 13, 79, 80, 87, 138, 194, 195

United States Agency for International Development (USAID), agricultural programs of, 219

United States Department of Agriculture, 8

Upland rice: definition and culture of, 19; impact on national yield levels, 142-144; percentage of, 143; weed control for, 55

Urea, application of: in large granules and briquets, 41

U.S.A. *See* United States

USAID. *See* United States Agency for International Development

U.S.S.R. *See* Soviet Union

Varietal improvement. *See* Rice breeding

Vietnam, 3, 20, 93, 150, 159, 165

Vitamins, in rice grain, 10-11

WARDA. *See* West African Rice Development Association

Water management, for modern varieties, 42-43. *See also* Irrigation systems

Water supplies, national surveys of, 167-168

Waxy rice. *See* Glutinous rice

Weed control, 53-56, 172, 176; influence of, on nitrogen response, 55; for rainfed rice, 158-159

West Africa. *See* Africa

West African Rice Development Association (WARDA), 218

White-backed planthopper, 46

Whorl maggot, 46, 49, 157

Wimberly, James E., 65 n

World Bank. *See* International Bank for Reconstruction and Development. *See also* International Development Association

World rice prices. *See* Rice prices, world

Xanthomonas oryzae. See Bacterial blight

Xanthomonas translucens. See Bacterial streak

Yield. *See* Rice yields

Yield potentials: of irrigated rice, 202; of rainfed rice, 202-203; studies of, by International Rice Research Institute, 25-29

Zaire, 7

Zinc, 40, 62, 153, 174, 197

Printed and bound by CPI Group (UK) Ltd, Croydon, CR0 4YY

23/10/2024

01778244-0002